SpringerBriefs in Geography

SpringerBriefs in Geography presents concise summaries of cutting-edge research and practical applications across the fields of physical, environmental and human geography. It publishes compact refereed monographs under the editorial supervision of an international advisory board with the aim to publish 8 to 12 weeks after acceptance. Volumes are compact, 50 to 125 pages, with a clear focus. The series covers a range of content from professional to academic such as: timely reports of state-of-the art analytical techniques, bridges between new research results, snapshots of hot and/or emerging topics, elaborated thesis, literature reviews, and in-depth case studies.

The scope of the series spans the entire field of geography, with a view to significantly advance research. The character of the series is international and multidisciplinary and will include research areas such as: GIS/cartography, remote sensing, geographical education, geospatial analysis, techniques and modeling, landscape/regional and urban planning, economic geography, housing and the built environment, and quantitative geography. Volumes in this series may analyze past, present and/or future trends, as well as their determinants and consequences. Both solicited and unsolicited manuscripts are considered for publication in this series.

SpringerBriefs in Geography will be of interest to a wide range of individuals with interests in physical, environmental and human geography as well as for researchers from allied disciplines.

Jürgen Schweikart · Jordi Martí-Henneberg ·
Jonas Pieper · Mateu Morillas-Torné ·
Bennet Schulte

Regional European Geographic Information System (REGIS), 1870–2020

A GIS Database for Thematic Mapping and Analysis

Jürgen Schweikart 
Berlin University of Applied Sciences
and Technology
Berlin, Germany

Jonas Pieper
Berlin University of Applied Sciences
and Technology
Berlin, Germany

Bennet Schulte
Berlin University of Applied Sciences
and Technology
Berlin, Germany

Jordi Martí-Henneberg
University of Lleida
Lleida, Spain

Mateu Morillas-Torné
University of Lleida
Lleida, Spain

ISSN 2211-4165　　　　　　　　ISSN 2211-4173　(electronic)
SpringerBriefs in Geography
ISBN 978-3-032-14931-2　　　　ISBN 978-3-032-14932-9　(eBook)
https://doi.org/10.1007/978-3-032-14932-9

This Springer imprint is published by the registered company Springer Nature Switzerland AG
The registered company address is: Gewerbestrasse 11, 6330 Cham, Switzerland

If disposing of this product, please recycle the paper.

Preface

This publication is the culmination of a project whose origins date back to the 1990s. Its primary aim was to develop *Regional European Geographic Information System* (REGIS), a reliable, transparently structured, and openly accessible database that enables comparative analysis of Europe's territorial evolution over extended historical periods.

A scientific approach always took center stage. Data collection, processing, and documentation were carried out in strict accordance with clearly defined methodological standards, ensuring the highest level of transparency. Historical sources were systematically analyzed, boundary changes were critically assessed, and cartographic representations were designed to be accessible and useful for researchers and students alike.

In addition, *Regional European Thematic Maps* (REThM) was created as a second database based on a generalized subset of the REGIS data. REThM specifically addresses the needs of small-scale mapping, facilitating the production of historical overview and comparative maps with reduced geometric complexity.

Together, REGIS and ReTHM serve a broad academic audience—particularly historians and geographers—and offer significant potential to related fields that rely on historical regional data, especially from periods when administrative boundaries differed from those of today. These resources may be valuable tools for spatially anchored research for economic historians, anthropologists, historical geographers, naturalists, and heritage studies, among other fields. We hope that, as well as supporting individual research projects, this work will stimulate discussion on the opportunities and limitations of digital historical data in spatial studies.

Our sincere gratitude goes to *The Mannheim Centre for European Social Research* (MZES) at the University of Mannheim, led in the project's early phase by Prof. Dr. Peter Flora, who championed this initiative from the start and provided invaluable support. We would like to give our greatest thanks to Franz Kraus, the head of the MZES Research Archive. His extraordinary leadership, unwavering commitment, meticulous attention to detail and tireless efforts in sourcing, verifying and organizing historical materials were essential in bringing this vision to life.

We are likewise grateful to the members of the Department of Geography at the University of Lleida in Spain, whose committed collaboration significantly contributed to the project's success. We also thank the former Beuth University of Applied Sciences Berlin—now the Berliner Hochschule für Technik—for its financial support, which was vital to the project's development.

Finally, we wish to acknowledge the many students who served as research assistants or contributed through their theses; without their enthusiastic participation and substantial input, this work would not have been possible. Our thanks also go to all those whose contributions, though not named here, supported the project in various roles and phases.

The research done by Jordi Martí-Henneberg was possible thanks to grants of the Interreg-Poctefa "Niu-Link," AGAUR (2019FI_B1 00127), and the Erasmus+ Jean Monnet Chair (**101239206, "Cross-border areas"**)—Centre of Excellence at the University of Lleida.

Berlin, Germany   Jürgen Schweikart
Lleida, Spain   Jordi Martí-Henneberg
Berlin, Germany   Jonas Pieper
Lleida, Spain   Mateu Morillas-Torné
Berlin, Germany   Bennet Schulte

Competing Interests The authors have no competing interests to declare that are relevant to the content of this manuscript.

Contents

1 Introduction 1
    References 6

2 Methodological Approach 9
    2.1   Capturing the Border Geometries 9
    2.2   Technological Approach for Data Processing 9
    2.3   Geospatial Data Format 11
    2.4   Definition of Change 13
    2.5   Data Sources 14
    2.6   Tracing Boundary Changes 15
    2.7   Limitations 17
    2.8   Generalization from REGIS to REThM 18
    2.9   The Need for Generalization 19
    2.10 Possibilities of Automated Procedures 20
    2.11 Cartographic Manual Generalization 21
    2.12 The Results 22
    References 24

3 Structure of the REGIS Dataset 27
    3.1   Introduction 27
    3.2   European Sovereign States and Temporary Regions 28
    3.3   Level Structure of the Territorial Organization 31
    3.4   Periodization of the Shapefiles 36
    3.5   Attributes of the GeoPackage Layers 39
    3.6   Datasets for the Whole of Europe by Decades 41
    3.7   Overview of the Download Structure 47
    References 51

4 Use of the REGIS and REThM Data 53
    4.1   Working with GeoPackage and Thematic Data in QGIS 54
         4.1.1 Opening and Displaying GeoPackage Layers 54
         4.1.2 Linking External Thematic Data 55

4.1.3   Creating Thematic Maps from Linked Data 59
4.2   Use Cases and Application Examples 61
    4.2.1   Population Change in Scandinavia 63
    4.2.2   Deriving a Collection of Historical Thematic Maps 65
    4.2.3   Infant Mortality in Europe 71
    4.2.4   Change Detection: Physician Density in Germany 72
References 78

5 Conclusion 81

Appendix 85

List of Figures

Fig. 2.1 REGIS layer in QGIS with connection between the polygon geometry and the associated attribute data ... 12

Fig. 2.2 Georeferenced sheet from Scobel (1906) with digitized GeoPackage layers ... 15

Fig. 2.3 Example of the digitization of a boundary line from Scobel (1906) ... 17

Fig. 2.4 REGIS with different scales ... 20

Fig. 2.5 Basic procedures during generalization (according to Hake et al., 2002) ... 22

Fig. 2.6 Comparison of the data on the scale of 1:17 million ... 23

Fig. 3.1 Changes in sovereign state borders between 1871 and 2020 ... 29

Fig. 3.2 List of all layers in the GeoPackage "REGIS_Romania.gpkg" ... 38

Fig. 3.3 Three different border situations in Romania at level 0 ... 39

Fig. 3.4 Excerpt from border changes in Romania 1960 on level 1 ... 39

Fig. 3.5 Excerpt showing the geometry and attribute information of the "REGIS_Germany_L3_1874-1876" layer ... 40

Fig. 3.6 Structure of ID and its relation to parent_id demonstrated using units from the layer "REGIS_Germany_L3_1874-1876" ... 41

Fig. 3.7 Availability of regional subdivisions in REGIS/REThM, based on the 10-year interval 1870: 1870–75 ... 44

Fig. 3.8 Availability of regional subdivisions in REGIS/REThM, based on the 10-year interval 1920: 1920–28 ... 45

Fig. 3.9 Availability of regional subdivisions in REGIS/REThM, based on the 10-year interval 1940: 1935–40 ... 46

Fig. 3.10 Availability of regional subdivisions in REGIS/REThM, based on the 10-year interval 1950: 1950–52 ... 47

Fig. 3.11 National and regional borders in Europe in 1870 (using REThM) ... 48

Fig. 3.12 National and regional borders in Europe in 2020 (using REThM) ... 49

Fig. 3.13   Overview of the file structure within the provided
             GeoPackages 50
Fig. 4.1   Step-by-step guides to opening GeoPackage layers in QGIS 55
Fig. 4.2   Example of overlaying two GeoPackage layers for Poland
             in QGIS 56
Fig. 4.3   Excerpt from a historical census document, Ireland, 1926
             (Central Statistics Office, 2025) 58
Fig. 4.4   Step-by-step instructions on adding data from the PDF
             to a GeoPackage layer (Part 1) 58
Fig. 4.5   Step-by-step instructions on adding data from the PDF
             to a GeoPackage layer (Part 2) 59
Fig. 4.6   Step-by-step instructions on styling the map in QGIS 60
Fig. 4.7   Map with Layer Styling dialog 60
Fig. 4.8   Example of a fully developed thematic map with essential
             elements 62
Fig. 4.9   Map of population growth in Scandinavia 64
Fig. 4.10   All maps in order of their appearance in the publication
             (Marti-Henneberg, 2021) 66
Fig. 4.11   Country maps of Romania (first row) and historical maps
             (second row) 68
Fig. 4.12   Modified Lambert Azimuthal Equal Area projection
             with center 50°0′ N, 10°0′ E (right) 70
Fig. 4.13   Cartographic perfection of details 71
Fig. 4.14   Map of infant mortality rate in Europe, 1991 73
Fig. 4.15   Visualization of historical physician densities in Germany 74
Fig. 4.16   Calculating the mean physician density of 1884
             in the boundaries of 2000 75
Fig. 4.17   Mean physician densities of historical censuses
             in the boundaries of 2000 77

List of Tables

Table 2.1	Scale of (entire REGIS) map and size of printing	18
Table 2.2	Properties and application of the two datasets	23
Table 3.1	Sovereign states and the time periods of their representation in the data	32
Table 3.2	Temporary states and regions and the time periods of their representation in the data	33
Table 3.3	Administrative or statistical levels of sovereign states	35
Table 3.4	Administrative and statistical levels of the UK and its constituent countries in the dataset	37
Table 3.5	Temporary states and regions and the time periods of their representation	37
Table 3.6	Number of units at each level (L0, L1, and L2) REGIS/ RETHM for the whole of Europe	42
Table 3.7	Population threshold and description of the NUTS levels, according to Eurostat (*Source* Eurostat, 2015)	42
Table 4.1	Assignment of the gray levels to the objects displayed on the map	69

Chapter 1
Introduction

The territorial development of Europe between 1870 and 2020 was characterized by dynamic changes that had a profound impact on politics, the economy and society. During this period, the continent witnessed a multitude of transformative processes: the redrawing of borders through wars and treaties, the emergence and dissolution of states, as well as regional autonomy movements and supranational integration efforts. These developments are not only of historical significance but also form a crucial basis for understanding Europe's contemporary political and cultural landscape.

To comprehensively understand the complex historical and spatial dynamics that have shaped modern Europe, a robust foundation of diverse resources is essential. These include theoretical frameworks that provide interpretative lenses through which territorial developments can be analyzed and empirical data that substantiate and contextualize these theoretical insights. Theoretical models allow for a structured examination of processes such as state formation, border reconfiguration, and regional integration. Meanwhile, empirical evidence—derived from historical data, statistical analysis and cartographic sources—ensures that these conceptual understandings are rigorously grounded in observable phenomena. Together, these resources form a critical nexus for advancing scholarly inquiry into Europe's evolving territorial and societal structures.

Europe exhibits significant spatial disparities, not only in terms of economic and socioeconomic development but also in the frequency and nature of changes to administrative boundaries.

To illustrate this, two extremes can be highlighted: On the one hand, countries like Spain, whose provincial boundaries have remained almost unchanged since 1833, provide a stable basis for long-term analyses. On the other hand, countries like Poland or Germany have experienced frequent and sometimes dramatic border changes, particularly as a result of the geopolitical upheavals of the twentieth century. These contrasts highlight the challenges of data comparability and the analysis of territorial developments.

© The Author(s) 2026

J. Schweikart et al., *Regional European Geographic Information System (REGIS),*
1870–2020, SpringerBriefs in Geography, https://doi.org/10.1007/978-3-032-14932-9_1

To address these challenges effectively, it is essential to examine specific examples in greater detail. The boundaries of the provinces of Spain have remained stable since 1833; this permits the study of these territories over time using census data and economic indicators. In this context, several studies have examined the evolution of regional disparities over time, analyzing long-term trends in economic inequality and their implications for regional development and integration (Diez-Minguela et al., 2018; Rosés et al., 2010; Tirado et al., 2016). The evolution of territorial imbalances is a subject that is very relevant for the territorial integration of a country, but this can only be analyzed when the territorial units are comparable over long periods of time. For this very reason, this type of analysis cannot be carried out—and that is the second example—in countries such as Germany and Poland, where border changes were dramatic and frequent until the end of the Second World War. The same occurred in most European countries. Significant border changes occur across all levels of spatial aggregation, affecting both national and regional units. At the national level, these changes manifest through the creation of new states, territorial reconfigurations, or geopolitical shifts. At the regional level, they include the adjustment of administrative units, the reorganization of internal boundaries, or the reassignment of jurisdictions. These processes highlight the complex and dynamic nature of territorial organization, with implications that extend beyond politics to encompass socioeconomic and cultural dimensions. A comprehensive analysis of these developments requires consideration of both macroscopic changes at the state level and microscopic dynamics on the subnational scale. Such an analysis is only feasible with reliable geometric data, as these provide an accurate and consistent representation of boundaries and spatial structures.

Theoretical considerations of territorial and spatial dynamics have been extensively explored and elaborated in a wide range of academic books (Dittel et al., 2024; Flora, 2013; Martí-Henneberg, 2023; Medeiros, 2017; Rokkan, 1999). Empirical studies have also addressed these issues, often supporting theoretical frameworks with practical examples and data. However, the application of a geographical approach, using the capabilities of Geographic Information Systems (GIS), has consistently been hampered by the lack of adequate digital geometric datasets. This limitation applies both to the temporal dimension, where datasets are often incomplete or inconsistent over time, and to the spatial dimension, where sufficient regional granularity is often lacking.

The lack of digital datasets with sufficient spatial and temporal depth is addressed by the publication of this database—called Regional European Geographic Information System (REGIS)—which aims to overcome these limitations. By providing a comprehensive and systematically structured database of geometric data with unique spatial and temporal granularity at the same time, this resource establishes the digital foundation necessary for detailed analyses of Europe's territorial evolution. It facilitates empirical investigations and opens up new possibilities for differentiated spatial studies that were previously limited by the lack of consistent geodata.

Our digital map data serve as a tool that enables the analysis of European regions, either as a whole or by subareas, using real boundaries for any period from 1870 to 2020. The system is designed for two scales, REGIS and a derived database called

Regional European Thematic Maps (REThM). For broader thematic maps covering larger regions or entire countries (small-scale maps), it is expected that they will also be maintained by the statistical institutions of the EU in the future. However, for detailed maps focusing on smaller areas and local administrative units (large-scale maps), there is currently no structure in place to ensure the continuation of the data.

This tool is of particular interest for several reasons. First, it will facilitate the analysis of the transformation of state and regional boundaries through the convulsive history of Europe. This topic has already been studied by several authors, both at the state level (Christopher, 1999) and at the regional level (Martí-Henneberg, 2023).

Secondly, our map series will contribute to a future Historical GIS that incorporates and combines different types of regional data, such as demographic, economic, transport allocation, and environmental indicators. In fact, it will be possible to transform any type of data into regional indicators that can then be displayed and studied in the format of real historical maps. An example of this would be the aggregation of the urban population of a given region; this would be obtained from the sum of the populations of the municipalities that have more than a certain population threshold. Because cities have permanent coordinates, we will be able to calculate and visualize the level of urbanization of all regions from 1870 to the present day, regardless of their changing shape. However, it is regrettable that this approach cannot be fully addressed with the data provided by REGIS. While the system offers valuable support, achieving a complete solution would necessitate independent research into municipal boundaries, a task that is both time-intensive and resource-demanding. Another significant example, in a completely different field, would be to study remote sensing data (based on aerial photographs and/or satellite images) and, for example, add the area of built-up areas within a given region. The number of possibilities is therefore enormous, since digital map series with all the boundary changes have not been available until now, and they will now facilitate developments in the fields of didactics and research that can consider the historical dimension.

Thirdly, the potential for diachronic analyses based on our digital maps will be considerable, as they allow for the examination of how regional and territorial phenomena evolve over time. These analyses can reveal long-term trends, historical development, and shifts in spatial patterns. However, it is important not to overlook the challenges associated with collecting the necessary data, as historical datasets, especially those involving changes in administrative boundaries, can be complex and difficult to obtain consistently. While it may seem relatively simple to represent information about a region at different points in time, analyzing the development of a phenomenon over time in spatial terms becomes much more complex, especially when the boundaries of the regions in question change. The core challenge lies in the fact that indicators are typically represented as average values (such as those used by Rosés and Wolf (2018), which are applied to the entire area. Typical examples of such indicators include population density or gross domestic product (GDP), which are calculated for the whole region. However, these indicators are rarely evenly distributed across the entire region; they are often concentrated in specific areas, such as population in settlements or GDP in economically active centers. When comparing regions whose borders have changed over time, this can lead to significant distortions. In such cases, one solution would be to rely on fine-grained data. However, these are often difficult to obtain. Furthermore, in many cases, even the fine-grained boundaries themselves have changed, presenting a dilemma that is frequently not easily resolved.

To make use of geometrical data effectively, attribute data are essential. One solution is to utilize census data, which has a long tradition in Europe. Apart from the counting of parts of society or simply the creation of tables, the first recognized population censuses were conducted in Sweden 1749 (Sköld, 2004) and Norway 1769 (Thorvaldsen, 2004), setting a precedent for systematic data collection. By the twentieth century, almost all European countries had established regular census cycles, typically conducted every 5–10 years, focused on systematically collecting demographic, social, and economic data. The compilation of censuses in Europe by Goyer and Draaijer (1992) serves as a comprehensive resource.

Population censuses are the most reliable source of data that we have for this kind of investigation, as they subdivide space into separate territorial units, and provide multiple, periodic snapshots, normally at regular intervals of 10 years. National statistical offices and research institutes have frequently neglected the historical dimension of such data and have often failed to compile continuous records that consider administrative border changes—at both the regional and municipal levels—and their impact on population data. Notable exceptions include the efforts of Nordic national statistical offices (Enflo et al., 2018) and a research project carried out in Spain. The latter (Fanch et al. 2013) produced a unified historical dataset of population changes, adapted to the current municipal map, composed of 8100 district local units.

Given the pivotal importance of censuses, the development of the database placed a particular emphasis on the temporal benchmarks provided by census data. These offer structured, periodic snapshots that serve as a solid foundation for analyzing historical and spatial trends.

This type of work has been the basis for national Historical GIS programs. The point of reference in Europe is the National Historical GIS, although the one carried out in the USA (Fitch & Ruggles, 2003) has undoubtedly been the most developed. In Europe, some attempts have been made to develop such programs at both the national and pan-European levels, but only for specific aspects, such as: economic development (Rosés et al., 2010), urbanization, and the development of railway networks (Martí-Romero et al., 2021). These National Historical GIS covers the period for which modern censuses have been compiled by European states: from the mid-nineteenth century to the present. Examples include Britain (Gregory, 2002), Belgium (De Moor & Wiedemann, 2001), the Netherlands (Boonstra et al., 2005), Germany (Kunz & Boehler, 2005), and France (Litvine et al., 2022). These countries have all made considerable progress in developing National Historical GISs in this form.

The further development of such Historical GISs will require the overcoming of two major obstacles. The first, and most fundamental, is the creation of digital maps of the type that we provide in this book. As already pointed out, this task consists in creating a GIS database of Europe including official regional boundaries in such a way that they correspond to census data. Having these maps will enable us to make comparisons across the whole continent at selected dates, but it will not allow us to directly explore changes over time, due to the previously mentioned problem of changing administrative boundaries. As mentioned above, the second challenge was to interpolate the census data with reference to a single set of administrative units to enable data to allow comparisons over time as well as space. In a previous work, we achieved this by creating a fully integrated GIS of European population data. This will now provide a valuable resource for a wide range of researchers who are interested in studying Europe's historical and modern population (Gregory, 2003; Logan et al., 2004).

To clarify terminology for the following chapters, it's important to distinguish between "levels"—which refer to the historical administrative hierarchies represented in the dataset—and "layers," the GIS-specific visual representations of those units. Unlike the EU's NUTS classification system, which is based on population size and policy planning needs, our levels are derived from historical administrative or statistical divisions used in censuses and statistical yearbooks. As such, they vary widely between countries. For instance, our Germany's level 1 (federal states) corresponds to NUTS-1, whereas our Croatia's level 1 (Županije) aligns with NUTS-3. Some countries, like Germany, include more detailed subdivisions (levels 2 and 3), while others do not. Our selection of levels is driven solely by historical structures and available historical statistics, not by modern classifications like NUTS, which in some cases (e.g., Austria's NUTS-3 regions) have no real administrative equivalent (see Chap. 3).

Unlike modern systems with standardized regional identifiers, historical units lacked consistent coding, making retroactive harmonization difficult. To address this, we created an internal ID structure that tracks administrative changes and parent–child relationships over time. However, since external data do not share this ID system, linkages rely on place name matching—a process complicated by historical spelling variations, language differences, and incomplete records (see Chap. 4).

The book is divided into five chapters, each addressing key aspects of the development and application of the database on Europe's administrative boundaries. Chapter 2 delves into the methodological foundations of the database creation, focusing on which border changes are included and the definitions underpinning these choices. Special attention is given to distinguishing between legally recognized boundaries and temporary ones, such as those established during wars, which are excluded from the data collection. Additionally, the chapter discusses the sources used, the processes for their integration, and the generalization of geometric data, particularly for small-scale map representations.

Chapter 3 explores the structuring of REGIS data. These are organized by national affiliation, administrative level, and temporal intervals, enabling precise and flexible analysis. The data foundation was derived from sources such as statistical yearbooks, and censuses. Based on this material, maps were created to trace border changes, area sizes, and the life cycles of administrative units. A precise definition of the study area is provided in Sect. 3.2.

Chapter 4 presents practical applications of the database, providing guidance even for users with no prior GIS experience. Rather than introducing QGIS, which is well-supported by numerous freely accessible resources, the focus will be on practical applications using the software. QGIS, a widely used open-source tool for geodata visualization and analysis, is continuously enhanced by a global community of developers. In this context, specific use cases will be explored, demonstrating how QGIS can be utilized for advanced spatial analysis and map creation, with a focus on its practical implementation within our project.

The final chapter reflects on the experiences gained throughout the project, offering conclusions about the work so far. It also outlines prospects for the use and further development of the database. Through this approach, the book serves as both a methodological foundation and a practical guide for researchers and practitioners.

At the core of this publication are the digital datasets that can be directly used for further research and applications. These materials are provided as supplementary resources accompanying Chap. 3.

This book is specifically targeted at advanced undergraduates, graduate students, and established researchers in geography, history, and related disciplines. It offers a unified GIS-based framework for investigating Europe's territorial evolution, while the accompanying database supports rigorous diachronic and comparative analyses of socioeconomic, demographic, and political developments across multiple spatial scales.

References

Boonstra, O. W. A., Doorn, P. K., & Schreven, L. (2005). *Towards a historical geographic information system for the Netherlands (HGIN)*. Reports on National Historical GIS Projects.

Christopher, A. J. (1999). *The atlas of states: Global change 1900–2000*. Wiley.

De Moor, M., & Wiedemann, T. (2001). Reconstructing Belgian territorial units and hierarchies: An example from Belgium. *History and Computing, 13*(1), 71–97.

Diez-Minguela, A., Martinez-Galarraga, J., & Tirado-Fabregat, D. A. (2018). *Regional inequality in Spain* (pp. 1860–2015). Springer International Publishing.

Dittel, J., Kühne, O., & Weber, F. (2024). A multi-perspective consideration of transformation processes in Europe and beyond. *Transformation processes in Europe and beyond: Perspectives for horizontal geographies* (pp. 3–19). Springer Fachmedien Wiesbaden.

Enflo, K., Alvarez-Palau, E., & Marti-Henneberg, J. (2018). Transportation and regional inequality: The impact of railways in the Nordic countries, 1860–1960. *Journal of Historical Geography, 62*, 51–70.

Flora, P. (2013). *European Regions. The territorial structure of Europe since 1870*. Palgrave Macmillan.

Franch, X., Martí-Henneberg, J., & Puig-Farré, J. (2013). Un análisis espacial de las pautas de crecimiento y concentración de la población a partir de series homogéneas: España (1877–2001). *Investigaciones Regionales= Journal of Regional Research* (25), 43–66.

Goyer, D. S., & Draaijer, G. E. (1992). *The handbook of national population censuses*, Europe.

Gregory, I. N. (2002). The accuracy of areal interpolation techniques: Standardising 19th and 20th century census data to allow long-term comparisons. *Computers, Environment and Urban Systems, 26*(4), 293–314.

Gregory, I. N (2003). *Arts, and humanities data service (England). History data service. A place in history: A guide to using GIS in historical research.* Oxbow.

Kunz, A., & Boehler, W. (2005). HGIS Germany: An information system on German states and territories from 1820 to 1914. *Historical Geography, 33*, 145–147.

Litvine, A., Gouj, H. E., Meyer, J., Starzec, A., Thévenin, T., Séguy, I. & Roy, D. (2022). Setting a standard for open and collaborative data acquisition for historical cartography: Digitizing the French État-major maps as a collaboration between the ANR COMMUNES and open historical maps. In *Proceedings of the 6th ACM SIGSPATIAL International Workshop on Geospatial Humanities* (pp. 1–7).

Logan, J. R., Zhang, W., & Xu, Z. (2024). Using public data to improve population estimates within consistent boundaries. In *The professional geographer* (pp. 1–10).

Martí-Henneberg, J. (2023). Boundary changes in Europe, 1850–2020. States, regions and data analysis. In *Creative ways to apply historical GIS: Promoting research and teaching about Europe* (pp. 1–11). Springer International Publishing.

Martí-Romero, J., San-José, A., & Martí-Henneberg, J. (2021). The radiality of the railway network in Spain during its early stages (1830–67): An assessment of its territorial coherence. *Social Science History, 45*(2), 363–389.

Medeiros, E. (2017). *Uncovering the territorial dimension of European Union cohesion policy.* Routledge.

Rokkan, S. (1999). *State formation, nation-building, and mass politics in Europe: The theory of Stein Rokkan: Based on his collected works.* Clarendon Press.

Rosés, J. R., Martínez-Galarraga, J., & Tirado, D. A. (2010). The upswing of regional income inequality in Spain (1860–1930). *Explorations in Economic History, 47*(2), 244–257.

Rosés, J. R. & Wolf, N. (2018). *The economic development of Europe's regions.* Taylor Francis.

Sköld, P. (2004). The birth of population statistics in Sweden. *The History of the Family, 9*(1), 5–21.

Thorvaldsen, G. (2004). The Norwegian census: An international and long-term perspective. *The History of the Family, 9*(1), 33–45.

Tirado, D. A., Díez-Minguela, A., & Martinez-Galarraga, J. (2016). Regional inequality and economic development in Spain, 1860–2010. *Journal of Historical Geography, 54*, 87–98.

Fitch, C. A., & Ruggles, S. (2003). Building the national historical geographic information system. *Historical Methods: A Journal of Quantitative and Interdisciplinary History, 36*(1), 41–51.

Chapter 2
Methodological Approach

2.1 Capturing the Border Geometries

The different tasks and applications in the field of geoinformation require different formats and programs, as no one option is suitable for all applications. There is a wide range of GIS software available to meet these needs, as well as a variety of proprietary and open spatial data formats.

Two key decisions must therefore be made at the outset: the choice of a geodata processing tool and the choice of a suitable geodata format. These decisions have to be made in light of different possible uses. The target groups are not necessarily experts in geoinformatics, but rather pupils, students, and scientists from different disciplines. Some may know little or not know how to work with spatial data. Others have already implemented projects with a strong spatial reference. It is therefore important to find a flexible solution that considers a variety of tools, not only in relation to the data, but also to further develop and add to the data collection presented.

The choice of technical solution allows the user to freely choose both the region and the time period and to combine these features as required, even if the entire database is not used. All in all, a solution is presented that makes it easy to use the database and open enough to add one's own data.

2.2 Technological Approach for Data Processing

A Geographic Information System (GIS) is a computer-based system consisting of hardware, software, data, and applications. It can be used to digitally record, store, manage, update, analyze, and model spatial data and present it alphanumerically and graphically (Bill, 2016; de Lange, 2006). A GIS can perform the two most important tasks of the project: constructing the geometric dataset and graphically representing the data as a map. Therefore, this project uses a small extract of the possibilities of

© The Author(s) 2026

J. Schweikart et al., *Regional European Geographic Information System (REGIS),*
1870–2020, SpringerBriefs in Geography, https://doi.org/10.1007/978-3-032-14932-9_2

a GIS. GIS is not just a data processing application, but also a scientific approach that addresses broader issues in geography, history, sociology, political science, and other fields (Longley et al., 2005).

In general, a GIS can manage models that utilize both vector and raster data, as well as their combinations, known as hybrid models. Raster data are irrelevant when building a system based on administrative boundaries; only a model based on vector data is useful. Spatial vector data are recorded as points, lines, and polygons. This application focuses on polygons because administrative units are made up of boundaries.

Geographical information systems were developed to process geographic data, including polygonal features. To achieve this, they require specific operational functionalities, such as data collection, geostatistical analysis, reporting, and visualization. For this application, the most crucial functions are:

- Creating georeferenced spatial vector data.
- Adding simple attribute data.
- Analyzing historical changes.
- Presenting data in thematic maps.

Geodata is typically categorized into geometric data and attribute data. Geodata is data that is associated with a specific location. This could include various types of information such as an address, a description, or a topological reference. The location is indicated by coordinates on the Earth's surface. To facilitate the use and development of the database by multiple users, a reference system appropriate for Europe needs to be selected. The attribute data in a GIS consist of non-spatial information associated with geographic features. These data describe the characteristics or properties of the spatial features and are typically stored in tabular format. Attribute data can include various types of information such as names, classifications, numeric values, dates, and descriptive text. They provide additional context and details about the geographic features represented in the GIS. For example, we could include demographic information such as the number of inhabitants, categorized by gender, within a polygon on our geographic map.

To effectively develop and utilize REGIS, a Geographic Information System (GIS) tool is indispensable. Further processing of REGIS data, such as integrating external attribute data and generating thematic maps, necessitates a software solution with expanded functionalities. There are various GIS software products available, including the well-known commercial tools within the ArcGIS product family from Esri, which provide comprehensive solutions for a wide range of applications. Other commercial GIS software products include GeoMedia from Hexagon Geospatial and MapInfo Pro from Precisely. However, these products can be expensive, even for basic license levels, making them less suitable for occasional users. Proprietary GIS tools, such as the ArcGIS product family, offer comprehensive functionality for a wide range of applications. They provide extensive support, including technical assistance and training materials. These tools seamlessly integrate with other software products and support standardized data exchange formats. However, proprietary GIS tools can be costly, particularly for full-feature packages or when used in large organizations with many users. Users are dependent on the provider for bug fixes, updates, and new feature releases. Additionally, these tools may lack the same level of flexibility as open-source alternatives.

Alternatively, there are various free and open-source desktop GIS products available today, each optimized for a specific purpose. Some of the most widely used software solutions with a broad range of functionalities include QGIS, SAGA, GRASS GIS, and gvSIG. These free and open-source options provide cost-effective solutions for users with varying needs and budgets. Open-source GIS products like QGIS offer several advantages. They are free to use, making them accessible to a wide range of users. Additionally, open-source software is flexible and customizable, allowing users to adapt it to their specific needs. QGIS benefits from a large community of users and developers, providing ample support and resources. Regular updates and ongoing development ensure that the software remains current and competitive.

However, open-source GIS products may have a smaller feature set compared to proprietary alternatives. Additionally, while there is community support available, it may not be as comprehensive or readily accessible as the support provided by commercial vendors.

Throughout the development of the REGIS database, a variety of GIS software tools were employed. Initially, data digitization and processing commenced in ArcMap (ESRI). The final stages of the project were executed using QGIS. Additionally, there was a shift in data format from ESRI shapefile to the open GeoPackage, a transition elaborated upon in the subsequent chapter.

2.3 Geospatial Data Format

According to Pons and Maso (2016), the format description documents database of the US Congress Library contained 334 formats in 2014, 34 of which have geospatial references. This number increases if metadata and symbolization instructions are contained in separate files (Kraak & Ormeling, 2003). It is important to identify the options that meet the requirements of the REGIS data. These requirements include wide distribution, ease of use without in-depth computer science knowledge, and continued support in the future.

Real-world object geometries can be stored as either vector or raster data. The main difference between raster and vector data is the way in which geographic features are represented. Raster data use a grid of cells to represent features, while vector data use points, lines, and polygons. Raster data are well-suited for continuous phenomena and analysis, while vector data are more suitable for discrete features and precise representation of spatial boundaries.

Administrative units in REGIS are represented as vector polygons. The effort to create them is worthwhile for several reasons:

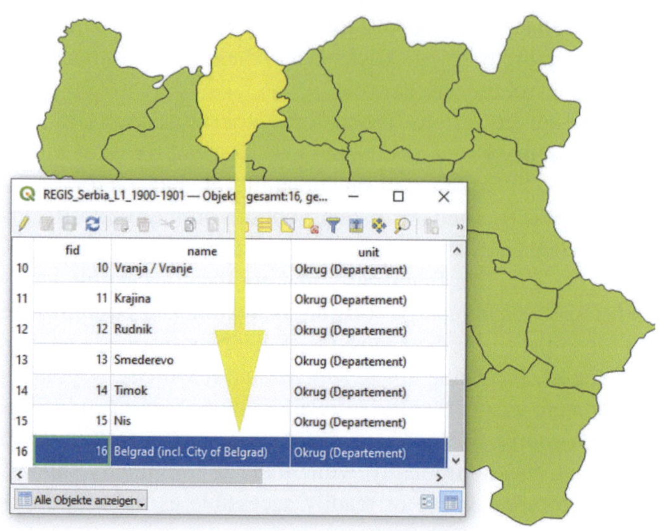

Fig. 2.1 REGIS layer in QGIS with connection between the polygon geometry and the associated attribute data

- There is no alternative to the arguments for a vector model. In summary, the advantages are as follows: Vector data allow for logical structuring and establishing object relationships between polygons and administrative units they represent. This enables linking geometries with associated or external attribute data. Figure 2.1 shows the link between the geometries of the administrative units of REGIS and the names of the units in the attribute table. When working with raster data, these options are severely limited.
- Vector data allow for the homogenization of regional administrative structures, as boundary lines in different map series can vary even if they represent the same boundary.
- Vector data have a lower data volume than raster data and require less computing time to display and examine.
- Many geoprocessing capabilities of GIS that are not available for raster data can be applied to vector data.
- The main advantage is that attribute data and statistics related to administrative units can be linked to vector data and visualized as thematic maps. This allows for clear and objective representation of data.

The ESRI shapefile format is the most commonly used format for storing and utilizing vector data in GIS. Originally developed as a proprietary file format for the GIS software ArcView from ESRI, it has become widespread due to ESRI's market-leading position. Although it was initially only supported by ESRI software, it can now be read by almost every commercial or free and open-source software. However, there are other modern formats for vector data that offer additional advantages in data management. One such format is the GeoPackage (*.gpkg), which was chosen to provide the vector data collected for REGIS. This is a relatively new, powerful, open, and manufacturer-independent geodata format that has been developed by the Open Geospatial Consortium (OGC) since 2014. It appears as a simple (geo)file, like the shapefile, but is a "one-file database." The GeoPackage format offers the benefits of a geodatabase while also providing the convenience of a single, easily exchangeable file. It allows for bundling of multiple vector polygon layers into a single GeoPackage file, making it an ideal choice for structuring REGIS data. The format is compatible with most commercial and open-source GIS programs available today. The QGIS community officially regards GeoPackage as the standard and preferred format due to its high level of functionality, making it easy to work with in QGIS.

2.4 Definition of Change

The aim of this work is to record all boundary changes that have occurred in European countries between 1870 and 2020, regardless of whether they were caused by conflicts between nation-states or by internal administrative reforms. A comparison of the historical changes in Spain and Germany highlights the different efforts required for each country. Spain has maintained almost complete stability of its external and internal borders, while Germany has undergone more than 25 structural changes over the same period (Pieper et al., 2007). Before collecting data, it is important to decide which changes should be included in the database. A pragmatic approach was adopted, considering the following aspects:

- The indicator chosen for updating the shapefiles is the percentage change in the area size. The aim is to record at least any area change of more than 10%, or less as a boundary change. The source used to control this is the censuses, which have been regularly carried out in Europe since the mid-nineteenth century and contain information on the area of the administrative units. The censuses also provide regionally differentiated information on the area sizes of the units. This ensures the validity of the GeoPackage layers.
- In addition, boundary changes should be verified using historical maps. Changes in unit size could also result from improvements in measurement methods.
- During times of war, border changes can be frequent due to the result of occupation and conquest. Changes that are not recognized under international law are usually not documented.

2.5 Data Sources

The starting point to produce dynamic geodata was the dataset "Regions - Nomenclature of Territorial Units for Statistics NUTS - March 1995," version NUTS V6 with the corresponding code "NUEC3MV6 1993," produced by EUROSTAT (Eurostat, 1995).[1] The NUTS boundaries with the scale of 1: 3 million were originally digitized from the paper map "Administrative map of the Community" published by the C.E.C. as part of the CORINE project. Subsequent modifications were made by EUROSTAT when changes were made to the NUTS nomenclature.

Starting from this dataset, which represents the 1995 boundary situation at NUTS-3 level for most countries, our work consisted in tracing all boundary changes back to earlier years and on to 2020. Based on this dataset, dynamic GeoPackage layers were digitized using information from administrative unit lifetime databases of the Mannheim Centre for European Social Research (MZES), Germany, and matching historical maps and atlases. A GeoPackage layer represents the boundary situation for one period at one administrative level. Most boundary changes could be recreated for all administrative levels by merging or splitting units based on the initial data set. For major changes, additional historical map sources had to be used. From these, individual boundaries were digitized.

Many sources are available as analog maps or atlases due to the length of time they have been processed. When digitizing administrative boundaries from analog maps, it is important to ensure accuracy, consistency, and adherence to scale. In this application, the digitization of historical map content focuses on boundaries within national units. The transformation into a GIS requires the lines to be generalized on a point basis. As a result, the line objects in the atlas maps are not completely congruent. Boundary transfer requires considerable creativity and time on the part of editors. The transfer is not always of complete boundary lines; in some cases, only sections of boundary lines are transferred.

Much of this data is based on scanned analog maps saved as raster data. The spatial reference was created by georeferencing in GIS, and the vector data were created by digitizing the borders of the scanned historical maps. In theory, a documentation of the historical administrative border changes could have been created simply as a raster data collection of scanned old maps. Several collections, including the David Rumsey Map Collection 2024, are accessible online. But using the data to create thematic maps or for further analyses would not have been possible.

Many cartographic documents and files have been examined in various archives and libraries around the world. A list of the main sources by country is given in the Appendix and will be made available with the online publication of the geodata. Three examples of important and frequently used historical sources are:

---

[1] Upon request, EUROSTAT confirmed in writing to the authors in June 2016 the free reuse of this dataset without payment or written license under the conditions stated in "© European Union, 1995 - today." They confirmed that the datasets do not fall under any of the exceptions listed there.

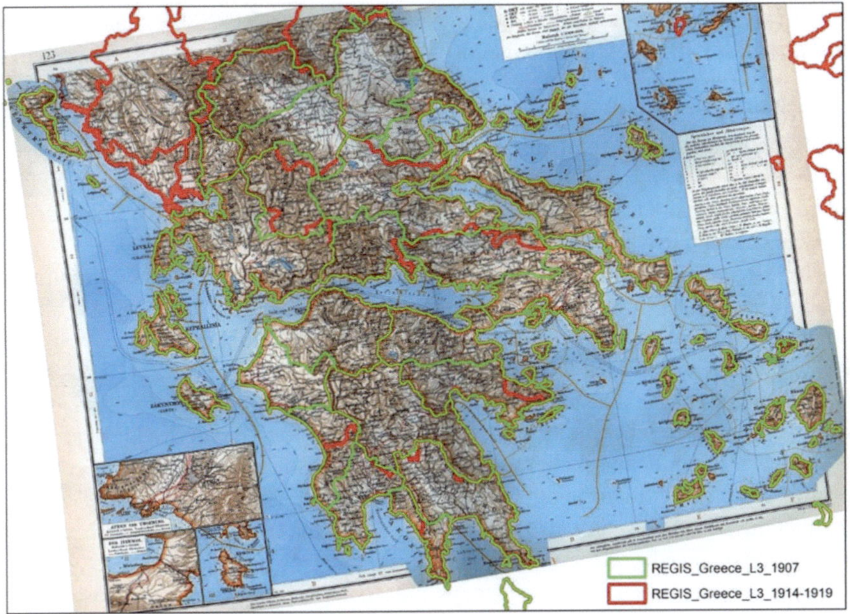

Fig. 2.2 Georeferenced sheet from Scobel (1906) with digitized GeoPackage layers

- Stieler's Handatlas über alle Theile der Erde und das Weltengebäude (Stieler & Vogel, 1873).
- Andrees Allgemeiner Handatlas, 5th Edition (Scobel, 1906).
- Atlante internazionale del Touring club italiano (Touring Club Italiano, 1929).

Figure 2.2 shows a georeferenced page from Andrees Allgemeiner Handatlas (Scobel, 1906), which was used to digitize border changes in Greece between 1907 and 1914. In addition to some atlases and maps that could be used for several countries, there are many maps that were used for specific countries and periods.

The appendix contains a comprehensive documentation of cartographic sources consulted, including notes on copyright status and licensing frameworks.

2.6 Tracing Boundary Changes

The REGIS database consists of dynamic GeoPackage layers for most European countries. The individual GeoPackage layers are based on the digitization of static benchmark maps, adapted to changes from a variety of different auxiliary maps. Each original map was previously standardized with respect to individual projections. This required the selection of control points in the original maps and reference data for georeferencing and transformation of the original maps into our standard projection, which is a Lambert Azimuthal Equal Area projection (ETRS89-extended/LAEA Europe [EPSG:3035]) which maintains the administrative units at their true relative sizes.

The external borders shown in REGIS, which are generally the coasts, always correspond exactly and at all administrative levels to the 1995 EUROSTAT dataset. Only the number of islands has been reduced as part of the generalization by removing islands with a diameter of less than 10 km. The aim is to improve the graph of the dataset according to the rules of good cartography.

The Mannheim Centre for European Social Research (MZES) is compiling a database based on the census reports of European countries since 1870 with the aim of identifying changes in administrative units. This collection of all published censuses in Europe is an indispensable and valuable source for the entire project. It has made it possible to create a database that contains information on when new administrative units were created, when units were dissolved, when they merged with other units or there were significant changes in their areas. In addition, the database contains information on the affiliation of units to units of higher administrative levels, and the fact that the sources usually also contain the area of the units means that changes in area can be deduced.

Using the georeferenced historic maps and a comparison with the database for the complete development cycle of the administrative units, the borders that have changed since 1995 (EUROSTAT dataset) are recorded and successively incorporated into the GeoPackage layers. The historical map sources are georeferenced using EUROSTAT boundaries and river courses, projected in the Lambert Azimuthal Equal Area. The historic map sources are georeferenced using the EUROSTAT boundaries and river courses in the Lambert Azimuthal Equal Area projection. Corresponding reference points that have not changed over the years are searched for in the GeoPackage layers and in the historic maps to be georeferenced. Once enough vertices have been found, a transformation method is applied for georeferencing. This process slightly alters the boundary geometries of the original map sources by distorting them to fit the target coordinate system.

Different digitization methods are available to deal with boundary changes between two time periods. In principle, boundaries can be redigitized because they did not exist before, or they can be deleted, for example as a result of the merging of two administrative units. However, there are often small changes that need to be made, such as moving or adding corner points. When digitizing original geometries, it is necessary to generalize, i.e., to simplify the boundaries of the original. To achieve a consistent quality, the software sets corner points at regular intervals and automatically connects them. The result is a generalized boundary (see Fig. 2.3).

When analyzing the changes that have occurred, two seem to be particularly common: Changes in the administrative organization of countries often involve merging smaller units into larger ones or dividing larger units into smaller ones. In these cases, the temporal dynamization of the GeoPackage layers could be achieved by simply merging or dividing units in GIS.

Fig. 2.3 Example of the digitization of a boundary line from Scobel (1906)

Merging units is also used to derive higher ranking regional units. In general, smaller administrative units of the lowest administrative level were digitized individually. However, care was taken to ensure that the boundaries of neighboring units were identical. In other words, they were digitized once and used in both units to define the polynomial. Higher administrative levels could then be derived by merging these smaller units in a GIS environment.

2.7 Limitations

The approach's liability lies not in its technical aspects, but rather in the real-world application of governmental boundaries and statistics in less organized states during the 19th and first half of the twentieth century.

To limit the number of changes due to possible micro border adjustments and to enable the verification of changes using historical or modern maps with large scales and uncertain accuracy, the area of change that creates a new condition in the data or in the GeoPackage layers must be restricted. It is therefore important to remember that changes in geometry are guaranteed to be considered if the area can be proven to have changed by more than 10% between two time periods. Minor changes were only considered if they could be reproduced in the map bases and area database without additional map material being researched. It should also be borne in mind that the quality of the sources used varies greatly. Large-scale maps do not give an accurate picture. They do not visualize data that is accurate in terms of surveying technology. The image is shaped by the cartographer who has already generalized the source data, especially when sources based on commercially manufactured products are used. Commercial publishing houses provide historical map sources that cover Europe. The accuracy and availability of high-quality geometrical sources vary geographically in terms of quality and reliability due to a focus on commercial sales markets. There is a significant difference between western and central Europe, on the one hand, and the Balkans or Russia, on the other hand. In particular, in the Balkan Peninsula, suitable map sources have occasionally been unavailable to reconstruct changes indicated by statistical data. Accordingly, short-term gaps in the temporal dynamization of the countries concerned are rare.

2.8   Generalization from REGIS to REThM

The European continent extends from the Scandinavian Peninsula to its southernmost point in Sicily for about 4000 km, and from west to east for about 5000 km. However, the project area dealt with in this paper is limited to a west–east extension of about 3350 km. In order to display this area on maps, it has to be scaled down. The visualization of the complete REGIS data (without Cyprus) in poster format (DIN A0) is possible in a suitable map projection with a scale of about 1:4.5 million, for its visualization in DIN A4 format a scale of at least 1:17 million is required (see Table 2.1). Clear and concise presentation of boundaries with such a small scale can only be achieved by simplifying them. This is known as cartographic generalization.

Table 2.1 Scale of (entire REGIS) map and size of printing

Scale	Printing size	DIN format
1: 1,000,000	400.0 cm × 335.0 cm	
1: 4,500,000	118.9 cm × 84.1 cm	A0
1: 6,500,000	84.1 cm × 59.4 cm	A1
1: 9,000,000	59.4 cm × 42.0 cm	A2
1: 12,500,000	42.0 cm × 29.7 cm	A3
1: 17,000,000	29.7 cm × 21.0 cm	A4

2.9 The Need for Generalization

Generalization is a scientific process aimed at adapting information to a map area reduced in terms of size and subsequently simplifying this information (Arnberger, 1997). Cartographic generalization has been systematically studied and developed in connection with the scale series of topographic maps. Maps with a relatively large scale, called primary scale, are used to derive map structures for maps with a smaller scale, called secondary scale (Bollmann, 2002). Many authors have identified cartographic generalization as the most intellectually demanding aspect and, in terms of technical feasibility, the automation of the generalization as the most technically challenging component of map production (João, 1998; McMaster & Shea, 1992). Generalization has several aims. Basically, the aim is to adapt the geographical structures and the information content to be derived from them to the reduced dimension of the outlines of the administrative boundaries of the map—and this is the sole objective for the present task. Relative to their actual size, objects can only be scaled down to a limited extent, i.e., to their minimum dimensions of cartographic symbology. This is due to the limitations of human visual perception. The output medium must be considered. There are clear differences between printed maps and the screen. For printed maps, values between 0.05 and 0.25 mm are given in the literature. For digital media, a value of slightly more than 1 pixel is recommended (Ledermann, 2022). When these minimum sizes are reached, the objects have to be scaled up in relation to the remaining content of the map. In this case, it is mainly the line thickness of the administrative boundaries that determines the level of detail to which the lines are still visible. This problem can only be solved by applying a variety of generalization techniques to simplify the administrative polygons of Europe.

The spatial data collected by Regional European GIS (REGIS) provide the user with a detailed equivalence map of the European regions and their boundaries, suitable for a wide range spatial analysis. Analyses with the scale class 1:1 million and above can be carried out at a high degree of accuracy. However, the data are not suitable for the production of a complete small-scale map of Europe (see Fig. 2.4). The aim is to create a way of modifying the maps in such a way that they can be used to display statistical data in small-scale thematic maps. As the focus is put on content and information, attributes such as horizontal accuracy, equivalence, and completeness may be neglected in favor of readability and uniqueness. Considering that the scale itself is the most important mathematical feature and the most important tool for classifying maps, this issue has to be considered as well within a GIS calculation. The scale with which the data have been digitalized is still a limiting factor for the degree of detail visible for the user (João, 1998). Generalization measures must refer to a certain map scale and then be displayed on the screen in a selectable exaggeration.

Figure 2.4 emphasizes the need for multiple map bases, adapted to the output scale. If the fjord coast of western Norway at 1:3 million seems appropriate, the same situation will result in an unsatisfactory map at 1:16 million. A similar situation can be seen in the Greek islands. This leads to the need for at least one additional

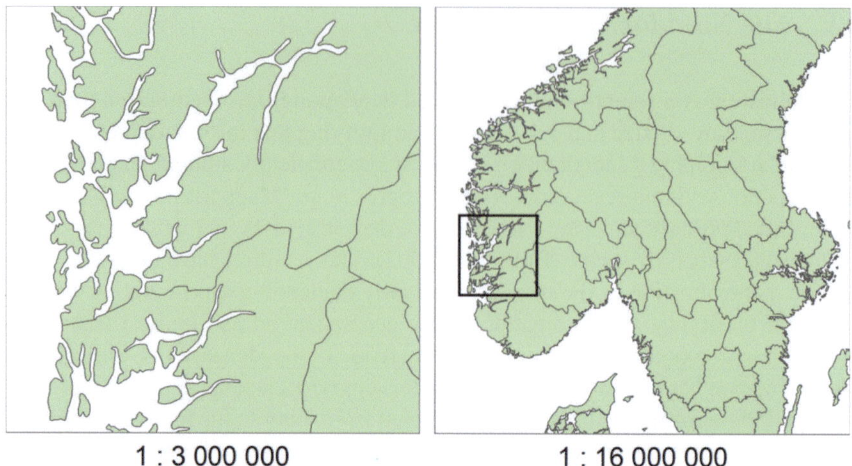

Fig. 2.4 REGIS with different scales

map base. This second dataset has been developed on the basis of REGIS under the name Regional European Thematic Maps (REThM). The aim is to create a base suitable for target scales between 1:10 million and 1:25 million. In the cartographic generalization of lines, it is important to retain essential geometric features such as direction, length, and connectivity, while reducing unnecessary detail to improve clarity and legibility when using smaller scales.

At the end of the process there are two map bases. One is REGIS for scales larger than 1:10 million, and the other is REThM for scales smaller than 1:10 million. These are mainly maps covering the whole of Europe or large parts of it. It is up to the user to choose between the two.

2.10 Possibilities of Automated Procedures

Line generalization techniques can also be applied to polygons. There are several methods, either based on elimination of vertices (simplification), or by deliberately modifying the course of the line using smoothing techniques (McMaster & Shea, 1992).

To generalize the REGIS data for a target scale of 1:17 million, tools from the most comprehensive and professional GIS software, like ArcGIS, were tested. There is a generalization toolbox that includes tools such as "Simplify Polygon," "Smooth Polygon," or "Aggregate Polygons," each with different algorithms that can be applied. A combination of these may produce satisfactory results for individual polygons, depending on the primary and target scale. In the context of the problem to be addressed, the generalization of complex polygon geometry should be carried out via a major scale conversion. This is a difficult task that cannot be performed by this simple tool or its combination. The original line characteristics are lost, and certain vertex constellations cause internal overlaps of line segments. If the parameter input is too low, the lines are not simplified enough for the target scale and gaps arise between the polygons. To achieve a sufficient degree of generalization, the parameter would have to be increased, resulting in even larger gaps.

None of the methods tested can give satisfactory results with such a large conversion between primary and target scale. The software products now available are aimed at complete solutions for topographic maps and are less suitable for pure polygon generalization, and they also require a high level of investment.

As the automated methods did not provide sufficient results, additional time-consuming post-processing of the results obtained was necessary. Automated procedures were tested without any claim to completeness of the available methods. None of the methods tested was able to deliver satisfactory results for the scale of the task. Automating map generalization has been an important research topic for decades, but the problem has not yet been fully solved (Courtial, 2024). In recent years, the introduction of deep learning (DL) and its inductive capabilities has given hope for further progress (Beglinger, 2023), so it is expected that problems such as the one presented here can be solved automatically in the future. In this case, the REGIS data were manually generalized cartographically.

2.11 Cartographic Manual Generalization

The methods of displacement, exaggeration, accentuation, and selection play an important role in the applied generalization of data to obtain and visualize the typical shapes of polygons. Particularly in coastal regions, these processes are still essential to obtain the characteristic shapes of the Norwegian fjords. Using a scale of 1:17 million, this cannot be achieved by vertex reduction and smoothing alone. With a completely new digitalization process, applicable to the 1:17 million scale and based on REGIS, the number of vertices can be significantly reduced by eliminating details. This method is also very time-consuming. However, experience and tests have shown that the challenge can be solved more efficiently by manual cartographic generalization.

Regarding the chosen procedure, the advantages of a holistic approach come to the fore. It is advantageous that all the necessary generalization procedures can be carried out in a single step, leading to an overall and inimitable result. The basic difference between digital and manual generalization is that manual generalization is a holistic process in terms of its view and design. The computer, on the other hand, operates with a limited logic, depending on the predefined algorithms (McMaster & Shea, 1992).

The process to be treated is a structural generalization process. It is based on a primary map that is transformed into a secondary map by means of cartographic generalization. The structure, and with it the graphical appearance of the polygons, is modified. Figure 2.5 shows the basic procedures (Hake et al., 2002).

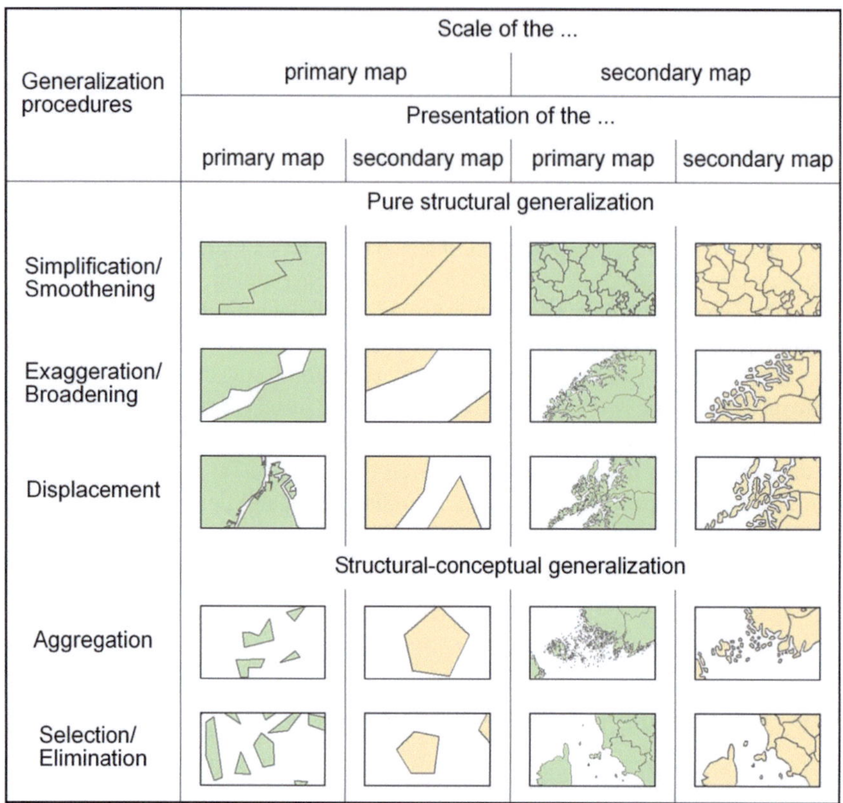

Fig. 2.5 Basic procedures during generalization (according to Hake et al., 2002)

Simplification, and therefore the smoothing of lines, seems to be the most important and most widely used technique. The exaggeration and widening procedures were used to map fjords, but also to represent smaller administrative units, e.g., in Switzerland or in Germany before 1945. Shifting was necessary in coastal regions to properly represent the islands. The aggregation method was also mainly used for mapping islands, e.g., in the Finnish "Åland Islands," where numerous islands cannot be mapped individually on a small scale. Selection and elimination have also played an important role in the generalization of islands and fjords in coastal regions.

2.12   The Results

Figure 2.6 shows a comparison between a section of the original REGIS data and the secondary product produced by generalization for thematic cartography (REThM) with the target scale of 1:17 million. This provides the user with a database that highlights characteristic structures. On this basis, the statistical data of the European regions can be displayed on a small-scale map. A comparative analysis covering several countries or even the whole of Europe on a single map sheet facilitates interpretation.

The two products REGIS and REThM are suitable for different applications. It is important that the user chooses the correct data according to the application. If the REThM data are used to calculate areas, accurate results cannot be expected due to the effects of generalization. Table 2.2 shows the characteristics of the products and their possible applications.

The REGIS data are best suited for use with a scale of about 1:5 million. Cartographic generalization is carried out to obtain a map of Europe with a target scale of 1:17 million. The result (REThM) is intended for thematic content representation. It does not have the structural properties of REGIS, i.e., it is not concretely representative. Users familiar with GIS as a tool for geographical analysis are aware of

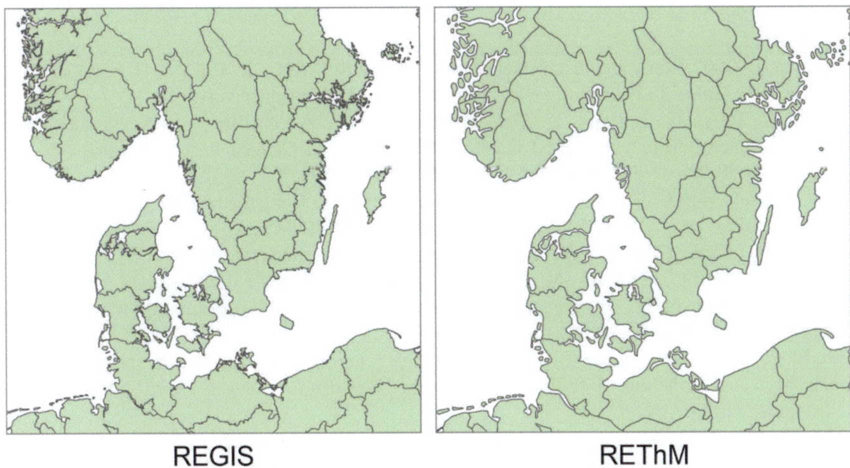

Fig. 2.6 Comparison of the data on the scale of 1:17 million

Table 2.2 Properties and application of the two datasets

	REGIS	REThM
Properties	• High accuracy • Countries can be mapped individually • Lines with high vertices density (fjord coasts) difficult to map	• Lower accuracy • High recall value of boundaries also with small scales • Details of boundaries and coastal lines are lost
Application	• In the scale range of 1:1 up to 1:10 million • Spatial analysis • Large-scale mapping	• In the scale range of 1:10 up to 1:25 million • Thematic mapping • Small-scale mapping

the limiting factors when using generalized datasets, which on the other hand also have advantageous features. With the appropriate data, many categories of thematic maps can be produced, showing the whole of Europe on a small scale. It is possible to reduce the map of Europe to a scale of 1:25 million. At the same time, all the possibilities of a GIS can be used. The map can be supplemented with additional georeferenced data whose accuracy is within the same scale range. To a certain extent, it is also possible to use specific sections of Europe for the production of thematic maps.

References

Arnberger, E. (1997). *Thematische Kartographie* (4th edn). Westermann, Braunschweig.

Bill, R. (2016). *Fundamentals of geographic information systems* (Vol. 6). Wichmann.

Beglinger, N. (2023). *Exploring Deep Learning for deformative operators in vector-based cartographic road generalization.* University of Zurich, Mathematisch-naturwissenschaftliche Fakultät.

Bollmann, J. (2002) Kartographische Generalisierung. In Bollmann, J., & Koch, G. (Ed.), *Lexikon der Kartographie und Geomatik* (Vol. 2). Spektrum Akademischer Verlag.

Courtial, A., Touya, G., & Zhang, X. (2024). DeepMapScaler: A workflow of deep neural networks for the generation of generalised maps. *Cartography and Geographic Information Science, 51*(1), 41–59. https://doi.org/10.1080/15230406.2023.2267419

David Rumsey Map Collection. (2022). David Rumsey Map Center, Stanford Libraries. Online (05.05.2025): https://www.davidrumsey.com/luna/servlet/RUMSEY~8~1

de Lange, N. (2006). *Geoinformatik in Theorie und Praxis*, Berlin (2nd edn).

Eurostat. (Ed.). (1995). *Regions—Nomenclature of territorial units for statistics nuts*, March 1995.

Hake, G., Grünreich, D., & Meng, L. (2002). *Kartographie: Visualisierung raumzeitlicher Informationen* (8th ed.). Walter de Gruyter.

João, E. M. (1998). *Causes and consequences of map generalisation.* Taylor & Francis Ltd.

Kraak, M. J., & Ormeling, F. (2003). *Cartography: Visualization of geospatial data* (2nd ed.). Longman Group.

Ledermann, F. (2023). Minimum dimensions for cartographic symbology—History, rationale and relevance in the digital age. *International Journal of Cartography, 9*(2), 319–341. https://doi.org/10.1080/23729333.2023.2165218

Longley, P., Goodchild, M., Maguire, D., & Rhind, D. (2005). *Geographic information systems and science* (2nd ed.). Wiley.

McMaster, R. B., & Shea, K. S. (1992). *Generalization in digital cartography.* Association of American Geographers.

Pieper, J., Schweikart, J., & Kraus, F. (2007). Generation and visualization of historical-administrative borders of Europe (1870–2000). In Tzschaschel, S., Wild, H., & Lentz, S. (Ed.), *Visualization of the room. Making cards—The power of cards*, forum ifl(6), 161–172.

Pons, X., & Maso, J. (2016). A comprehensive open package format for preservation and distribution of geospatial data and metadata. *Computers and Geosciences, 97*, 89–97. https://doi.org/10.1016/j.cageo.2016.09.001

QGIS Documentation. (2024). Online (05.05.2025): https://www.qgis.org/en/docs/index.html

QGIS Download. (2024). Online (05.05.2025): https://www.qgis.org/en/site/forusers/download.html

Scobel, A. (1906). *Andree's general hand atlas* (5th edn). Velhagen & Klasing.
Stieler, A., & Vogel, C. (1873). *Stieler's hand atlas on all parts of the earth and on the world building*. Composite: Germany. Justus Perthes, Gotha.
Touring Club Italiano. (1929). *Atlante internazionale del Touring club italiano*, Milano.

Chapter 3
Structure of the REGIS Dataset

3.1 Introduction

To enable the user to access the data as individually as possible, the GeoPackages of the two levels of generalization (REGIS and REThM) are each organized according to three features which are related to each other: The national affiliation and the administrative level refer to the time period mentioned.

The relevant information for defining the administrative levels and determining their units and duration of existence is based on the work of the research archive *Eurodata* of the *Mannheim Center for European Social Research* (MZES). One outcome of this work was the provision of data and additional information that are necessary for the construction of a dynamic map series based on the territorial units of European sovereign states.

Data were collected on the evolution of political-administrative levels, which included information on their number, designation, and political constitution. For the main levels, information was collected on all their units (entities), including their relations to superordinate entities as well as about the duration of their existence and changes in membership of superordinate entities. The data show the life history of the system levels and their associated territorial units and formed the basis for the search for suitable map material for digitization. In addition, contextual information on the changes in national and internal borders and the area size of the political-administrative units was also provided.

The information for this database was mainly collected from statistical yearbooks and censuses. Census data, typically collected every 10 years in most European countries, have primarily been used, but supplemented with statistical yearbook data,

---

Supplementary Information The online version contains supplementary material available at https://doi.org/10.1007/978-3-032-14932-9_3.

J. Schweikart et al., *Regional European Geographic Information System (REGIS), 1870–2020*, SpringerBriefs in Geography, https://doi.org/10.1007/978-3-032-14932-9_3

which were generally published annually, for intermediate periods where territorial changes exceeding 5% occurred between censuses. With the establishment of statistical offices in the mid to late-nineteenth century, the publication of statistical yearbooks soon became common practice, and reporting on political-administrative divisions became a standard part of the yearbooks. Censuses are an extremely important source for describing and analyzing historical social change: Comparing the data from the statistical yearbooks with those from the censuses is essential, especially since information gaps in the statistical yearbooks can often be filled by census publications.

3.2 European Sovereign States and Temporary Regions

To establish a unified database spanning 150 years, it is first necessary to define the external boundaries of Europe. However, there is no single, universally accepted answer to the question, "What constitutes Europe?" Europe is not a homogeneous entity, but rather a dynamic space characterized by cultural diversity and shared historical foundations. From a geographical perspective, Europe is the second-smallest continent and is commonly defined as the westernmost part of the Eurasian landmass. Its boundaries include the Arctic Ocean to the north, the Atlantic Ocean to the west, the Mediterranean Sea to the south, and the Ural Mountains, Ural River, and Caspian Sea to the east. This physical delineation encompasses a variety of climatic zones, ranging from the tundra of the north to the Mediterranean climate of the south. From a political perspective, Europe comprises numerous nation-states as well as supranational entities such as the European Union (EU). However, not all European countries are EU members. As a cultural and economic space, Europe has been shaped by a long and diverse history. It is defined less by natural borders and more by cultural, linguistic, and religious similarities and differences. Notably, the region embodies values such as cultural integration, multilingualism, and the preservation of cultural heritage (Gebhardt et al., 2013; Triandafyllidou & Gropas, 2023).

For the purposes of the present database, the definition of Europe's spatial boundaries has been approached pragmatically, considering the project's specific objectives and data availability, depending on the context.

Russia and Turkey are considered as non-European countries because geographically, most of their territories are located outside the European continent. In the context of this dataset, the spatial framework for representing the administrative boundaries of European countries is based on the criterion of de facto independent statehood. Nevertheless, certain exceptions apply—most notably for territories that were historically integrated into the Russian Empire or the Soviet Union.

Finland is one such exception. Although it did not gain full independence from Russia until the final stage of the First World War (1917/1918), it is represented in the dataset from 1870, covering a total of 48 years prior to its formal independence. This representation is justified by Finland's special administrative status within the

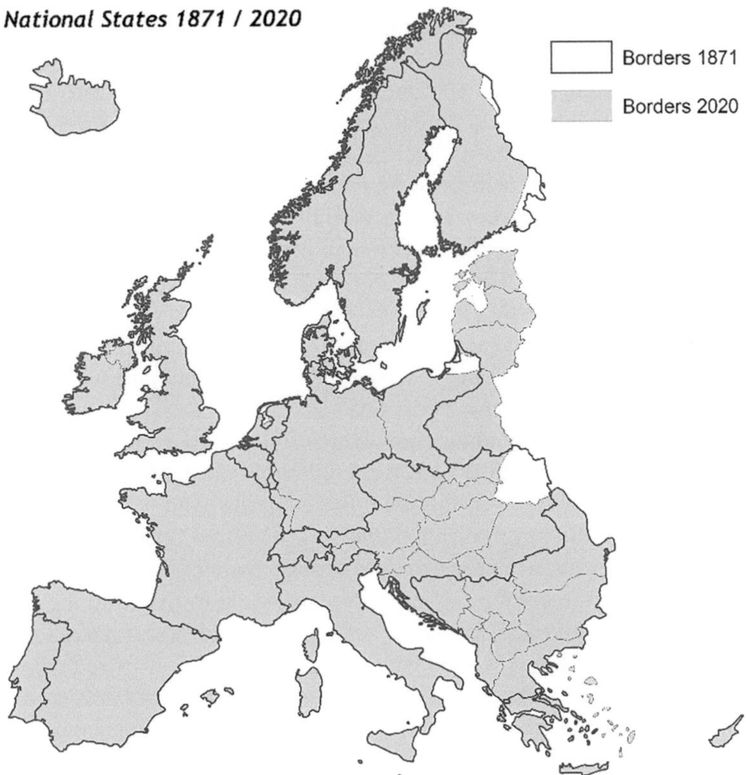

Fig. 3.1 Changes in sovereign state borders between 1871 and 2020

Russian Empire, where it retained a degree of autonomy and institutional separation. The inclusion of Finland in the dataset for this period allows for a more nuanced and historically accurate reflection of territorial and political developments (see Fig. 3.1).

The Baltic states—Estonia, Latvia and Lithuania—also gained independence from Russia during the final stage of the First World War (1917/1920). However, unlike Finland, they did not form a unified or clearly delineated administrative entity within the Russian Empire prior to their independence. As such, they are only included in the dataset during their independent inter-war period (1920–1940). Following their incorporation into the Soviet Union at the beginning of the Second World War, they cease to be represented in the dataset as independent entities. They are again represented in the data after regaining sovereignty in the context of the dissolution of the Soviet Union in 1990/1991.

In contrast, Ukraine, Belarus, and Moldova gained full independence for the first time within the timeframe covered by this project during the collapse of the Soviet Union. As these territories were not independent entities before 1990, they are not included in the dataset prior to that date. Furthermore, unlike other former Soviet republics, these three countries are deliberately excluded from the dataset even after 1990, as a specific exception within the project's methodological framework.

Instead, parts of present-day Ukraine are depicted as former Austro-Hungarian territory, aligning with historical borders and acknowledging their different administrative and cultural history before their incorporation into the Soviet Union.

South of the Black Sea, the area shown is bounded by the current border between Greece and Bulgaria with Turkey. Before World War I, large parts of the Balkans up to this border are shown as part of the Ottoman Empire. Cyprus since independence and Iceland are included in the dataset. Figure 3.1 compares the boundaries of the area shown in 1871 and 2020.

While state borders on land are governed by international, mostly bilateral agreements, maritime borders along the coast are defined in accordance with the United Nations Convention on the Law of the Sea (UNCLOS). Coastal state boundaries are determined by the so-called baseline, which follows the natural contour of the coastline. Territorial waters, up to 12 nautical miles from the baseline, are considered part of the sovereign territory of the state. For countries with a coastline, national borders extend into the sea, typically following maritime boundaries defined by international agreements. However, these maritime borders are not represented in the dataset. Instead, the coastline is used as a substitute to delineate the extent of the land area. This approach ensures a consistent and clear cartographic visualization of national territories while maintaining a focus on land-based administrative boundaries. These coastlines represent the natural boundary between land and water, providing a realistic and accurate representation of geographical boundaries. Coastal lines also serve as an important reference for orientation and facilitate the understanding of a country's geographical location, which is particularly evident in Norway's distinctive coastline. Coastlines have been used as reference points to define country boundaries in both datasets (REGIS and REThM).

Europe's coastlines have changed over the period, both because of natural processes and human activity. These include erosion and sedimentation, as well as human influences such as land reclamation projects—for example, in countries such as the Netherlands and Denmark—port construction and coastal defenses. It is difficult to quantify these changes over the observation period. In particular, land reclamation in the Netherlands has created about 2,000 km^2 of new land since the nineteenth century, through projects such as the diking of the Zuiderzee (now the Ijsselmeer) and the creation of polders. These effects were considered for the Netherlands. All other coastlines remain unchanged throughout the entire period for the dataset. The representation of the coastlines corresponds to the situation in 1993, based on the "NUEC3MV6 1993" dataset, produced by EUROSTAT (Eurostat, 1995).

All sovereign states that existed in the described geographical area between 1870 and 2020 are represented in the dataset as "Sovereign states." Table 3.1 lists all the sovereign states included in the dataset, together with the basic lifetimes represented by the geodata. Not all states presented here have necessarily been recognized as sovereign by all countries. For example, at the time of writing, Kosovo has not been recognized as a sovereign state by some European countries, including Spain, Greece, Romania, and Slovakia. The basically represented lifetimes are the times for which geodata are basically available in the datasets, regardless of their level structure or boundary changes during these times.

The aim of this project was to record and map the border configurations of sovereign and independent nation-states in Europe, together with their internal administrative structures, to provide a consistent and comparable spatial basis for the projection of numerical data. This geographical framework allows the integration of statistical or administrative data for most of the European continent from 1870 to 2020.

Periods characterized by rapid border changes or widespread lack of international recognition have been deliberately excluded. In such periods there is no single, stable geographical base on which statistical data could be uniformly projected. These so-called multi-perspective periods, often characterized by conflicting de jure and de facto borders, make national statistics incomparable because of the different territorial references between states. However, border changes due to war, occupation, or military conquest were not recorded in the dataset. This leads to intentional gaps in the geodata for the countries affected. In cases where post-war border demarcations were not immediately formalized through international agreements, the dataset only begins once a widely accepted territorial status has been established.

There are also some gaps in the lifetimes of the geodata, where no suitable cartographic material could be found for the digitization of border situations. Therefore, the years specified in the table do not always represent the actual lifespan of a sovereign state but rather the period within its existence for which geodata is available in the datasets. The years listed in Table 3.1 always refer to January 1 of the year in which the boundary situation represented by the geodata was definitively in effect.

Temporary semi-sovereign entities that did not receive full international recognition or were subject to the League of Nations are represented as "Temporary States and Regions" (see Table 3.2). This includes the European part of the Ottoman Empire, the present-day Balkans up to the Turkish border.

The datasets contain two files (GeoPackages) for each sovereign state and each temporary state or region represented. One GeoPackage for the more detailed boundary representation (REGIS) and one for the generalized boundary representation (REThM). The GeoPackage can be identified by the abbreviation in front of the country name, e.g., for Albania the two files are "REGIS_Albania.gpkg" and "REThM_Albania.gpkg."

3.3 Level Structure of the Territorial Organization

One of the principal objectives of our geodata is to facilitate the display of historical data within the context of its historical boundaries. The levels of regional subdivision are therefore based on the historical levels used in censuses or statistical yearbooks. The objective is not to standardize spatial-statistical or administrative levels, as is the case with the NUTS classification of the European Union (European Union, 2022).

Table 3.1 Sovereign states and the time periods of their representation in the data

Nation	Basically depicted lifetime at level 0 (national border)
Albania	1921–1926, 1927–2020
Andorra	1870–2020
Austria	1922–1938, 1946–2020
Austria-Hungary	1875–1914
Belgium	1870–2020
Bosnia and Herzegovina	1993–2020
Bulgaria	1880–1912, 1923–1940, 1947–2020
Croatia	1992–2020
Cyprus	1870–2020
Czechia	1993–2020
Czechoslovakia	1921–1939, 1946–1992
Denmark	1870–2020
Estonia	1919–1940, 1991–2020
Finland	1870–2020
France	1871–1914, 1920–1939, 1945–2020
GDR	1949–1990
Germany	1871–1918, 1921–1938, 1949–2020
Greece	1870–1919, 1928–1940, 1951–2020
Hungary	1920–1939, 1949–2020
Iceland	1870–2020
Ireland	1922–2020
Italy	1872–1915, 1921–1941, 1948–2020
Kosovo	2009–2020
Latvia	1920–1939, 1992–2020
Liechtenstein	1870–2020
Lithuania	1921–1939, 1991–2020
Luxembourg	1870–2020
(North) Macedonia	1992–2020
Malta	1870–2020
Monaco	1870–2020
Montenegro	1879–1913, 2007–2020
Netherlands	1870–1939, 1946–1956
Norway	1870–2020
Poland	1922–1939, 1946–2020
Portugal	1870–2020
Romania	1879–1913, 1926–1939, 1952–2020
San Marino	1870–2020

(continued)

Table 3.1 (continued)

Nation	Basically depicted lifetime at level 0 (national border)
Serbia	1879–1912, 2007–2020
Serbia and Montenegro	2004–2006
Slovakia	1993–2020
Slovenia	1992–2020
Spain	1870–2020
Sweden	1870–2020
Switzerland	1870–2020
UK	1871–2020
Vatican City	1930–2020
Yugoslavia	1920–1939, 1948–2003

Table 3.2 Temporary states and regions and the time periods of their representation in the data

Temporary state or region	Basically depicted lifetime at level 0 (national border)
Crete	1899–1913
Eastern Rumelia	1878–1885
Free City of Danzig	1921–1939
Free State of Fiume	1921–1923
Free Territory of Trieste	1948–1954
Italian Aegean Islands	1924–1943
Memel Territory	1920–1923
Part of Ottoman Empire	1870–1913
Saarland	1920–1935, 1948–1956

It is not possible to make comparisons between the levels of the different countries in terms of size or number of inhabitants. In the most recent year, 2020, included in the dataset, for instance, the federal states of Germany are depicted as level 1, which aligns with the NUTS-1 level according to the NUTS classification. Conversely, in Croatia, the Županija is illustrated as level 1, which corresponds to the NUTS-3 level in the present era. The German GeoPackages contain additional, more detailed levels (2 and 3), which are not included in the Croatian packages. The selection of administrative or statistical levels shown should always be viewed in the context of historical development and the potential availability of statistical data. This approach is based on the work of the Mannheim Center for European Social Research (Quick, 1994).

In the GeoPackages, small states such as Monaco are displayed solely with their outer borders, without additional levels of regional organization. The outer borders of a country are designated as level 0. In countries with a larger population, at least level 1 is represented, with regional divisions that were also used for population censuses. In other countries, level 2 is also included, while a few countries have level 3.

To improve clarity, we need to distinguish between the terms "level" and "layer." "Level" refers to the different administrative levels represented in the geospatial data. "Layer" is a GIS-specific term that refers to a feature class displayed as a separate layer within a Geographic Information System.

A GeoPackage for a sovereign state contains a varying number of layers corresponding to different administrative levels. The number of levels was explained in the previous paragraph. The number of layers per administrative level depends on the number of boundary changes that occurred. Each unique border configuration is represented by its own layer. If no boundary changes took place at one level, there is only one layer; if there were ten boundary changes, there would be eleven layers (one for the initial state plus ten for the changes).

Table 3.3 presents only the sovereign states with their respective levels, which are included in the datasets with additional regional levels situated below the national borders (level 0). If the entirety of a given level is represented by a singular designation for the units in question, no annual figures are provided. If the units of a given level have undergone alteration over the course of their lifespan, the periods of validity are indicated. Additionally, instances where a level did not exist for the entirety of a sovereign state's lifespan, or where the map bases for its digitization could not be located for the entirety of that lifespan, are also noted. In some sovereign states, the units at a given level are designated differently. In the event of a multitude of disparate designations, the most significant ones are presented, with "& more" indicating the existence of additional designations at this level. If a given level and its constituent units correspond to a level within the current NUTS classification system, the corresponding NUTS level is indicated in parentheses.

The UK presents a unique case due to the distinct regional subdivisions of its constituent countries, especially when considering historical contexts. Northern Ireland, England, Ireland (prior to 1922), Scotland, and Wales each have their own administrative structures. As a result, these regions are represented on separate layers in the two GeoPackages for the UK. The specific layers are identified by including the name of the respective constituent country in their designation (e.g., "REGIS_UK_England_L2_1971-1975"). Details of the mapped levels for each region are provided in Table 3.4.

Only two temporary states or regions are represented with an administrative level below the national level (see Table 3.5). Eastern Rumelia is mapped using the "Okrug" units, corresponding to level 2. This is because Bulgaria, which later absorbed Eastern Rumelia, already had a level 1 administrative unit above the Okrug. The Free Territory of Trieste is divided into two level 1 administrative zones: one administered by the Allies, later becoming part of Italy, and the other administered by Yugoslavia and later dissolved into it.

Table 3.3 Administrative or statistical levels of sovereign states

Sovereign state	Administrative level below the nation
Albania	**1** Prefekture 1921–1940; County *(NUTS-3)* 1947–2020 **2** District 1950–2000
Austria	**1** Bundesland *(NUTS-2)*
Austria-Hungary	**1** Kingdom and Empire **2** Crownland and more **3** Kerület, Megye, and more
Belgium	**1** Region *(NUTS-1)* **2** Province *(NUTS-2)*
Bosnia and Herzegovina	**1** Federation, Republic, and District **2** Canton
Bulgaria	**1** Rajoni *(NUTS-2)* 2004–2020 **2** Okrug 1880–1912 and 1923–1940 Oblast *(NUTS-3)* 1947–2020
Croatia	**1** Žubanija *(NUTS-3)*
Czechia	**1** Kraje *(NUTS-3)*
Czechoslovakia	**1** Zeme and more **2** Kraje 1949–1992
Denmark	**1** Amt (County) 1870–2007 Regioner *(NUTS-2)* 2008–2020
Estonia	**1** Maakond (County) 1919–1940 Maakondade *(NUTS-3)* 1991–2020
Finland	**1** Län (Province) **2** Maakunnat *(NUTS-3)* 1997–2020
France	**1** Région *(NUTS-2)* 1960–2020 **2** Département *(NUTS-3)*
GDR	**1** Land 1949–1952; Bezirk 1953–1990

Germany	**1** Kingdom, Principality, and more 1871–1918
	   Free/People 's State, and more 1921–1938
	   Federal/Free State *(NUTS-1)* 1949–2020
	2 Province, & more 1871–1918 & 1921–1938
	   District *(NUTS-2),* & more 1949–2020
	3 District and more 1871–1918 & 1921–1938

Greece	**1** Groups of Peripheries *(NUTS-1)* **2** Peripheries *(NUTS-2)* **3** Nomoi (Department) *(NUTS-3)*
Hungary	**1** Régió (Planning Region) 1999–2020 **2** Vármegye (County) 1920–1939 Megyék (County) *(NUTS-3)* 1949–2020
Iceland	**1** Landshlutar/Region 1919–2020
Ireland	**1** Historical Province **2** County (and County Borough) 1922–2000 (City and) County Council 2001–2020

(continued)

Table 3.3 (continued)

Sovereign state	Administrative level below the nation
Italy	**1** Compartimento 1872–1970 Regioni and Regione Autonoma *(NUTS-2)* 1971–2020 **2** Provincia and more *(NUTS-3)*

Latvia	**1** Apgabali 1925–1935
	   Rajons (district) 1992–2004
	   Statistiko regionu/statistical region*(NUTS-3)*
	2 Aprinkis

Lithuania	**1** Apskritys (County) *(NUTS-3)*
Netherlands	**1** Provincie *(NUTS-2)*
Norway	**1** Amt (County) 1870–1919 Fylke (County) *(NUTS-3)* 1920–2020
Poland	**1** Województwa *(NUTS-2)*
Portugal	**1** Distrito (District)

Romania	**1** Historical Province 1879–1939
	   Regunea 1952–1967
	   Regiuni (Region) *(NUTS-2)* 1982–2020
	2 Judete (County) *(NUTS-3)*

Serbia	**1** Okrug (Departement) 1879–912 Statistical Region 2007–2020 **2** Okrug (District) *(NUTS-3)* 2007–2020
Serbia and Montenegro	**1** Republika (Republic) **2** Statistical Region **3** Okrug (District)
Slovakia	**1** Kraj *(NUTS-3)*
Slovenia	**1** Statisticne regije (Statistical regions) *(NUTS-3)*
Spain	**1** Comunidad autónoma (Autonomous community) 1980–2020 *(NUTS-2)* **2** Provincia (Province) *(NUTS-3)*
Sweden	**1** Län (County) *(NUTS-3)*
Switzerland	**1** Kanton (Cantons) *(NUTS-3)*

Yugoslavia	**1** Oblast (Region) 1920–1929
	   Banovina (Region) 1930–1939
	   Socialisticna Republika 1948–1991
	   Republika (Republic) 1992–2003
	2 Soc. Autonomna Pokrajina (Auton. Province) 1948–1991
	3 Okrug (District) 1992–2003

3.4   Periodization of the Shapefiles

The periodization of the geodata for creating a dynamic and digital map series was developed using various information based on the work of the *Research Archive of the Mannheim Centre for European Social Research* (Eurodata). This includes:

Table 3.4 Administrative and statistical levels of the UK and its constituent countries in the dataset

	Constituent countries	Administrative level below the nation

UK	England	1 Kingdom 1871–2020
		2 Ancient County 1871–1888
		  Administrative County (incl. County Boroughs) 1889–1974
		  (non-)metropolitan county, and more (NUTS-3) 1975–2020

	Ireland	1 Kingdom 1871–1921 2 Historical province 1871–1921 3 (Civil) County (and County Borough) 1871–1921
	Northern Ireland	1 Constituent country 1922–2020 2 NUTS-3 Area 1974–2020 3 (Civil) County (and County Borough) 1922–1973

	Scotland	1 Kingdom 1871–2020
		2 Civil County 1871–197
		  Single- or Two-Tier system 1976–1996
		  Unitary Authority 1997–2020

	Wales	1 Constituent nation 1871–2020
		2 Ancient County 1871–1888;
		  Administrative County (and City Boroughs) 1889–1974
		  County (Two-Tier-System) 1975–1996
		  Unitary Authority 1997–2020

Table 3.5 Temporary states and regions and the time periods of their representation

Temporary state or region	Administrative level below the nation
Eastern Rumelia	2 Okrug 1878–1885
Free Territory of Trieste	1 Civil administration Zone 1948–1954

- Information on the development of the political-administrative organization of nation-states with details on their number of administrative regions, designation, and political constitution.
- Information on hierarchical relationships, such as father–child relationships between administrative units (e.g., provinces within a nation-state or municipalities within a province), along with details of their memberships, duration of existence, and changes over time. Contextual information on the change of borders and the change in political and administrative system levels.
- Information on the size and population figures of the political-administrative units. The data on changes in area sizes were used to periodize the validity period of each layer in the GeoPackages. All significant boundary changes that resulted in a change in a unit of more than five percent of the area of a unit should be recorded where possible.

The period during which the documented border situation existed is shown in the names of the various GeoPackage layers for each country. The years always refer to the situation on January 1 of the year indicated. Figure 3.2 shows an example

⊟ REGIS_Romania.gpkg
  > ▭ REGIS_Romania_L0_1879-1913
  > ▭ REGIS_Romania_L0_1926-1939
  > ▭ REGIS_Romania_L0_1952-2020
  > ▭ REGIS_Romania_L1_1879-1913
  > ▭ REGIS_Romania_L1_1926-1939
  > ▭ REGIS_Romania_L1_1952-1960
  > ▭ REGIS_Romania_L1_1961-1967
  > ▭ REGIS_Romania_L1_1982-2020
  > ▭ REGIS_Romania_L2_1879-1913
  > ▭ REGIS_Romania_L2_1926-1939
  > ▭ REGIS_Romania_L2_1968-1981
  > ▭ REGIS_Romania_L2_1982-1995
  > ▭ REGIS_Romania_L2_1996-2020

Fig. 3.2 List of all layers in the GeoPackage "REGIS_Romania.gpkg"

of the GeoPackage layers for Romania. If, for example, the years 1879–1913 are given, as in the first layer, then the border situations contained in the layer were valid on January 1 of the years given. However, the border situation shown may have already been created during the year 1878 (except on January 1). In the course of 1913, the border situation changed due to the Second Balkan War, so that the border situation no longer existed on January 1, 1914 (these wartime border situations are not represented by the layers).

After the First World War, the exact course of the borders within Romania could only be reconstructed with certainty using sources from 1925, so the inter-war period does not begin again until 1926 (January 1). This border situation at level 0 lasted until the beginning of the Second World War (1939). The period after the Second World War, which can be shown with certainty by suitable sources, only begins in 1952. This external border situation continues until today (2020). Figure 3.3 shows the three different external borders of Romania at level 0 for the period under study.

There have been additional changes at levels 1 and 2 in Romania as a result of major and minor administrative reforms. An example of a change at level 1, the former level of regions (Regiumea), in the course of 1960, is shown in Fig. 3.4. The number of units increased from 16 to 18. The cities (Oraşul) of Bucharest and Constanta were created at level 1 and are marked yellow in the figure shown. There have also been minor boundary changes between the existing units of Galati and Ploiesti, as shown in Fig. 3.4.

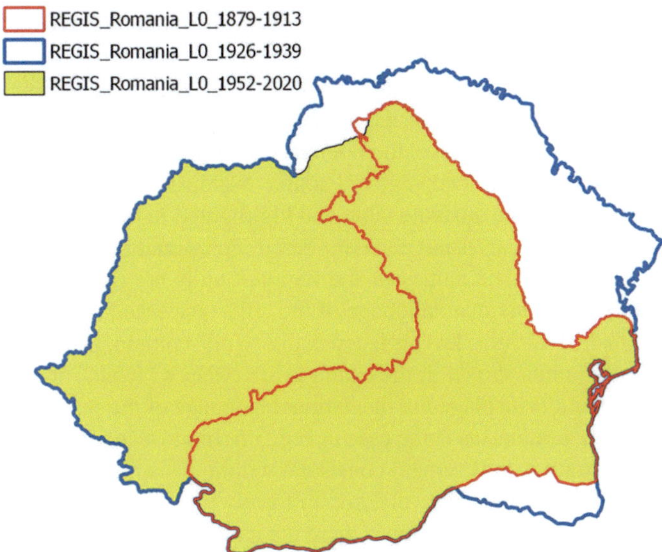

Fig. 3.3 Three different border situations in Romania at level 0

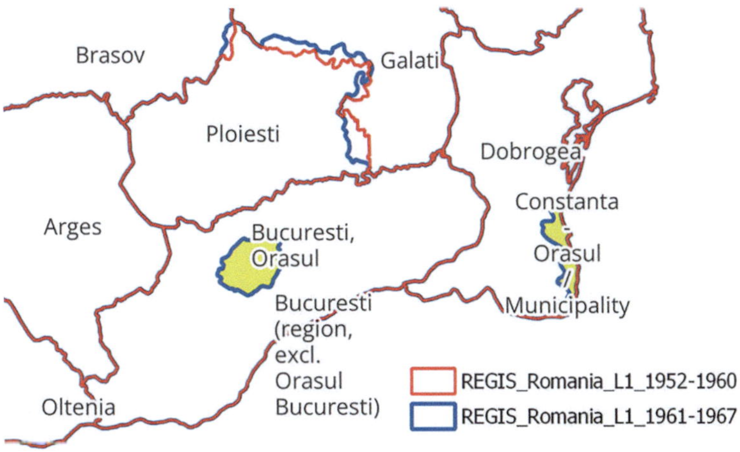

Fig. 3.4 Excerpt from border changes in Romania 1960 on level 1

3.5 Attributes of the GeoPackage Layers

The individual layers of each GeoPackage contain both spatial information about boundary geometries and associated attribute data. These layers can be opened and edited using any GIS software. In the GIS, the link between the attribute information and the polygon geometries is readily visible. Each spatial unit is represented by up to five entries in the attribute table.

Figure 3.5 provides an excerpt from the boundary geometries and attribute table of a layer from the GeoPackage representing the former German Empire. The example focuses on level 3, which captures the boundary configuration for the period 1874–1876. In this example, the two regional units "Waldeck" and "Neckarkreis" are specifically selected in the attribute table and highlighted in the corresponding map view. This provides a clear visual and data-based representation of their spatial and administrative context. The names of the regional units are listed in the "Name" column, usually in the national language of the units concerned.

The type of administrative level is listed in the "Unit" column. During the selected period, the level 3 units shown in the example are either a "Kreis" or a "Freistaat." The name of the unit is displayed in the national language of the selected country, in English or in both variants, as is the case in Fig. 3.5. In most countries, the designations for administrative units remain consistent within the same layer and level. For example, in France, all units at level 2 are uniformly referred to as "départements." However, there are notable exceptions where the names of administrative units vary, either within a single period or between different historical periods. This variation may be due to historical, political, or regional factors. The German dataset shown in Fig. 3.5 is an example of such a case. At level 3, administrative units are labeled with three different terms, including "Kreis/County," "Regierungsbezirk/District," and "Kreishauptmannschaft/Ancient County," among others not shown in the excerpt.

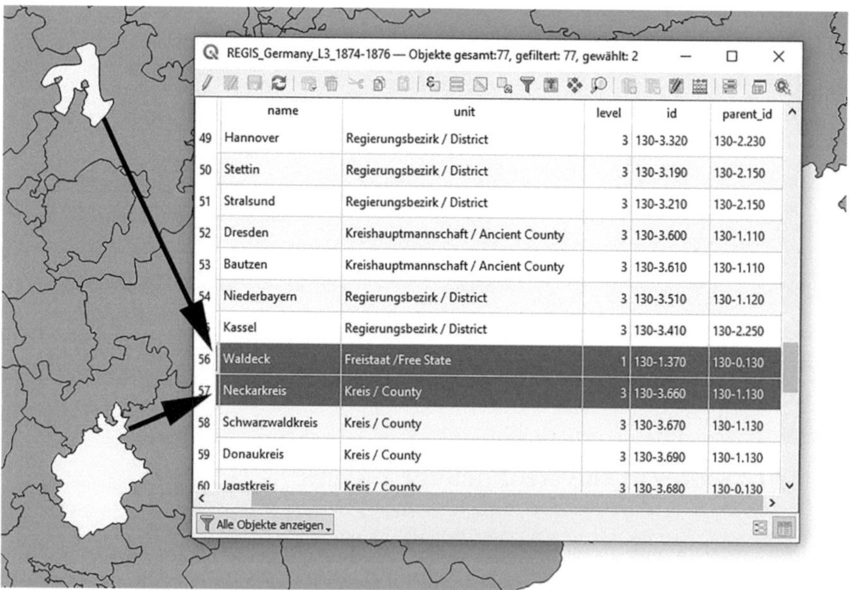

	name	unit	level	id	parent_id
49	Hannover	Regierungsbezirk / District	3	130-3.320	130-2.230
50	Stettin	Regierungsbezirk / District	3	130-3.190	130-2.150
51	Stralsund	Regierungsbezirk / District	3	130-3.210	130-2.150
52	Dresden	Kreishauptmannschaft / Ancient County	3	130-3.600	130-1.110
53	Bautzen	Kreishauptmannschaft / Ancient County	3	130-3.610	130-1.110
54	Niederbayern	Regierungsbezirk / District	3	130-3.510	130-1.120
	Kassel	Regierungsbezirk / District	3	130-3.410	130-2.250
56	Waldeck	Freistaat /Free State	1	130-1.370	130-0.130
57	Neckarkreis	Kreis / County	3	130-3.660	130-1.130
58	Schwarzwaldkreis	Kreis / County	3	130-3.670	130-1.130
59	Donaukreis	Kreis / County	3	130-3.690	130-1.130
60	Jaastkreis	Kreis / County	3	130-3.680	130-0.130

Fig. 3.5 Excerpt showing the geometry and attribute information of the "REGIS_Germany_L3_ 1874-1876" layer

	country code Germany	level indicator	unit identifier	complete ID	parent ID
level 0	130	0.	130 (Germany)	130-0.130	
level 1	130	1.	100 (example kingdom "Preußen")	130-1.100	130-0.130
level 2	130	2.	240 (example province "Westfalen")	130-2.240	130-1.100
level 3	130	3.	380 (example district "Münster")	130-3.380	130-2.240

Fig. 3.6 Structure of ID and its relation to parent_id demonstrated using units from the layer "REGIS_Germany_L3_1874-1876"

There are several other variations in addition to those visible in the extract, which further reflect the complexity of historical administrative structures.

In certain countries, higher-level administrative units are not subdivided further. For example, in the German Empire, the Hanseatic cities and free states held level 1 status and were not divided into smaller units (a status that persists for Hamburg, Bremen and Berlin today). The attribute table indicates the corresponding administrative level in the "Level" column. Figure 3.5 highlights the free state of Waldeck, which had level 1 status at the time and was not subdivided into levels 2 or 3. To ensure data continuity, higher-level units are repeated in the layers of lower levels if no units are defined for those levels.

The attribute table is further enhanced with unique identifiers (IDs) in the "id" column. These IDs consist of a three-digit prefix that identifies the country (e.g., "130" for Germany). The suffix indicates the administrative level (e.g., "3" for level 3) followed by a three-digit unit identifier (e.g., "660" for the "Neckarkreis"). The prefix and suffix are joined by a hyphen (see Fig. 3.5). The "parent_id" column assigns each unit to its corresponding higher-level unit. For example, as shown in Fig. 3.6, the level 3 district "Münster" belonged to the level 2 province "Westfalen," which, in turn, was part of the level 1 kingdom of "Preußen" (Prussia). At the time, Prussia constituted the largest part of the German Empire.

3.6 Datasets for the Whole of Europe by Decades

So far, we have presented and described the datasets that contain the national and regional boundaries of each sovereign state at specific points in time. In this section, we explain how these country-level datasets have been compiled into a single archive for each administrative level (L0, L1, and L2), and at two levels of generalization: REGIS and REThM. In other words, in addition to the individual datasets provided for each country, we also offer six comprehensive files, covering levels 0, 1, and 2 for both REGIS and REThM, each in GeoPackage format. These files contain harmonized boundaries for all of Europe, presented at ten-year intervals for each regional level.

Table 3.6 shows the number of territorial units included at each administrative level—states (L0), and regions (L1 and L2)—within the pan-European datasets. The regional classification used in this book follows a different logic and structure from Eurostat's NUTS classification (Nomenclature of Territorial Units for Statistics), although some overlaps exist (see Table 3.3). The NUTS framework distinguishes between national territories (NUTS-0), major socioeconomic regions (NUTS-1), regions relevant to cohesion policy (NUTS-2), and smaller units used for specialized analysis (NUTS-3) (Eurostat, 2015) (see Table 3.7).

A key issue with the NUTS system is that levels 1 through 3 are not strictly based on existing administrative divisions. Instead, they aim to define regions of broadly comparable population size across Europe. In some countries, NUTS regions closely align with administrative units; in others, they are aggregations or subdivisions of those units. Importantly, NUTS levels are not inherently comparable across national contexts, which has led researchers to adjust regional units to improve analytical consistency. For instance, several studies have combined NUTS-3 units in smaller countries to approximate the territorial scale of NUTS-2 regions in larger countries such as Germany or France (Gregory et al., 2009; Caruana-Galizia & Martí-Henneberg 2012; Martí-Henneberg, 2017).

Table 3.6 Number of units at each level (L0, L1, and L2) REGIS/RETHM for the whole of Europe

Level*	1870–1875	1880	1890	1900	1910	1920	1930	1938–1940
L0	28	32	32	33	33	42	42	41
L1	361	405	406	404	400	537	497	495
L2	653	725	729	718	735	855	830	825

Level*	1950–1952	1960	1970	1980	1990	2000	2010	2020
L0	39	37	37	37	37	46	45	45
L1	454	410	445	436	385	407	369	362
L2	747	771	783	782	763	931	902	889

* Level 0 corresponds to states, while levels 1 and 2 correspond to two regional levels

Table 3.7 Population threshold and description of the NUTS levels, according to Eurostat (*Source* Eurostat, 2015)

NUTS level	Minimum population	Maximum population	Description	Number of regions 2021
NUTS-0	–	–	States	37
NUTS-1	3,000,000	7,000,000	Major socioeconomic regions	104
NUTS-2	800,000	3,000,000	Basic regions for the application of regional policies	1283
NUTS-3	150,000	800,000	Small regions for specific diagnoses	1345

The classification used here—L0, L1, and L2—is designed to support geographic analysis and facilitate meaningful comparisons across regions. It consistently refers to administrative or statistical units used in historical censuses and official statistical publications.

The grouping of datasets for European countries enables seamless work across the entire continent, ensuring appropriate representation of data at the regional levels where they are available. The data could be provided to show annual variations, but a more instructive approach is to present the material through maps at 10-year intervals, aligned with years ending in 0 (e.g., 1870, 1880, 1890, …, 2000, 2010, 2020). However, exceptions arise when historical events allow for precise determination of borders at specific moments. For instance, negotiations following military conflicts often take years to finalize. Consequently, when compiling datasets for Europe for the most tumultuous years, in some cases, borders from nearby years were used by us to ensure consistency and accuracy in the dataset (see Tables 3.1 and 3.2 in the previous section).

For example, after the First World War (1914–1918) and the subsequent collapse of the Russian Empire (1917–1921), numerous successor conflicts reshaped eastern and southeastern Europe. The incorporation of Fiume (Rijeka) into Italy in 1924 marked the end of this period. As a result, the first round-number decade with broadly stable and internationally recognized borders across the region was 1930, deliberately leaving out 1920. Another example concerns the period of the Second World War. Between 1939 and 1945, statistical data referred to the administrative units imposed by the occupying authorities, while the government of the occupied nation-state continued to use the pre-war administrative divisions, which sometimes bore similar names but did not coincide geographically. Such discrepancies make direct comparisons impossible.

As a result, although there are temporal gaps in the geodata for certain countries due to wars, disputed sovereignty or delayed border settlements, a series of standardized and spatially comparable datasets are available for the beginning of each decade, with the following exceptions:

- 1870, due to the ongoing processes of national unification in Germany and Italy.
- 1920, due to the unresolved territorial consequences of the First World War and subsequent conflicts.
- 1940, due to the territorial upheavals of the Second World War already in progress.
- 1950, as a result of some post-war adjustments—such as the incorporation of the Dodecanese into Greece.

The maps on the following pages (see Figs. 3.7, 3.8, 3.9, and 3.10) indicate the first year within each 10-year interval (e.g., 1870) for which the REGIS or REThM dataset provides regional administrative subdivisions below the national level (Level 0). In cases such as Austria-Hungary—illustrated in Fig. 3.7—where the data begins in 1875, this does not imply that the national borders depicted were only established in that year. Rather, it reflects the earliest point for which reliable source material on internal regional divisions was available and processed within the REGIS/REThM framework. While the external borders of Austria-Hungary, as shown, were already in place from 1867 onward, the REGIS and REThM datasets provide detailed regional subdivisions at levels 1, 2, and 3 starting from 1875.

Figure 3.7: In 1870, Germany was in the final stage of unification under Prussia, culminating in the proclamation of the German Empire (Deutsches Reich) in January 1871 (Nagel, 2021). Similarly, the unification of Italy was completed with the annexation of Rome on September 20, 1870. However, the full administrative integration and the relocation of the capital from Florence to Rome were finalized in 1871, which is why we consider the Italian borders as of January 1872 (Bettoni, 2021). Reliable information on the internal borders of the Austro-Hungarian Empire was available from 1874 onward. Therefore, the geodata begins in 1875, although the state was established earlier. Accordingly, the borders for 1870 are representative of the period 1870–1875

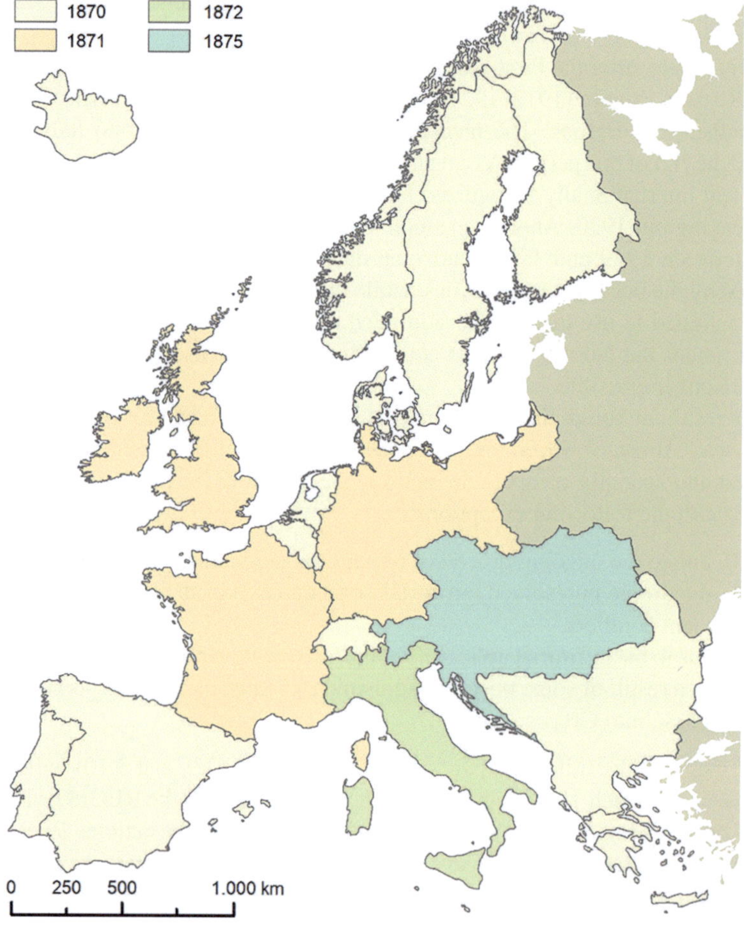

Fig. 3.7 Availability of regional subdivisions in REGIS/REThM, based on the 10-year interval 1870: 1870–75

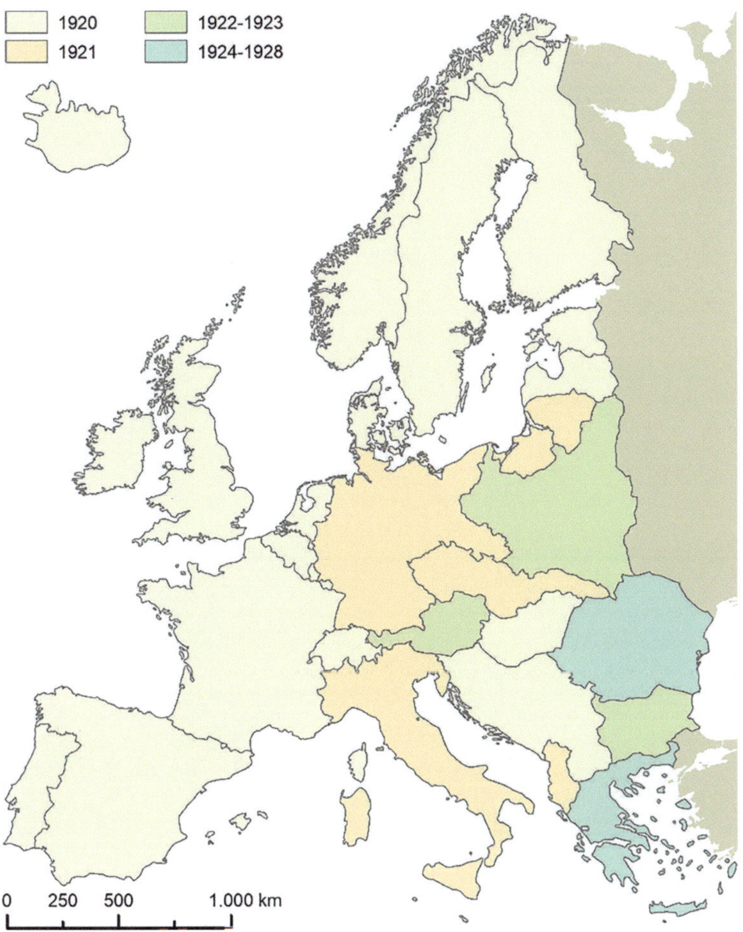

Fig. 3.8 Availability of regional subdivisions in REGIS/REThM, based on the 10-year interval 1920: 1920–28

Figure 3.8: A comparable situation occurred at the end of the First World War (1914–1918). Although the war ended with an armistice in 1918, the resulting political and territorial changes unfolded over several years (see Fig. 3.8). As a result of the Great War, several new states emerged at different points in time: Estonia declared independence in 1918 and was recognized in 1920. Hungary was established in 1918 following the dissolution of Austria-Hungary, while Latvia, Yugoslavia, and the Free State of Fiume were formed in 1919–1920. The Free City of Danzig was established in 1920 under the Treaty of Versailles. Other territorial changes involving Albania, Germany, Italy, Lithuania, Czechoslovakia, Austria, Poland, Bulgaria, Romania, and Greece occurred between 1920 and 1928

Figure 3.9: The Second World War (1939–1945) introduced significant political instability, both in the years preceding the war and in the immediate post-war period,

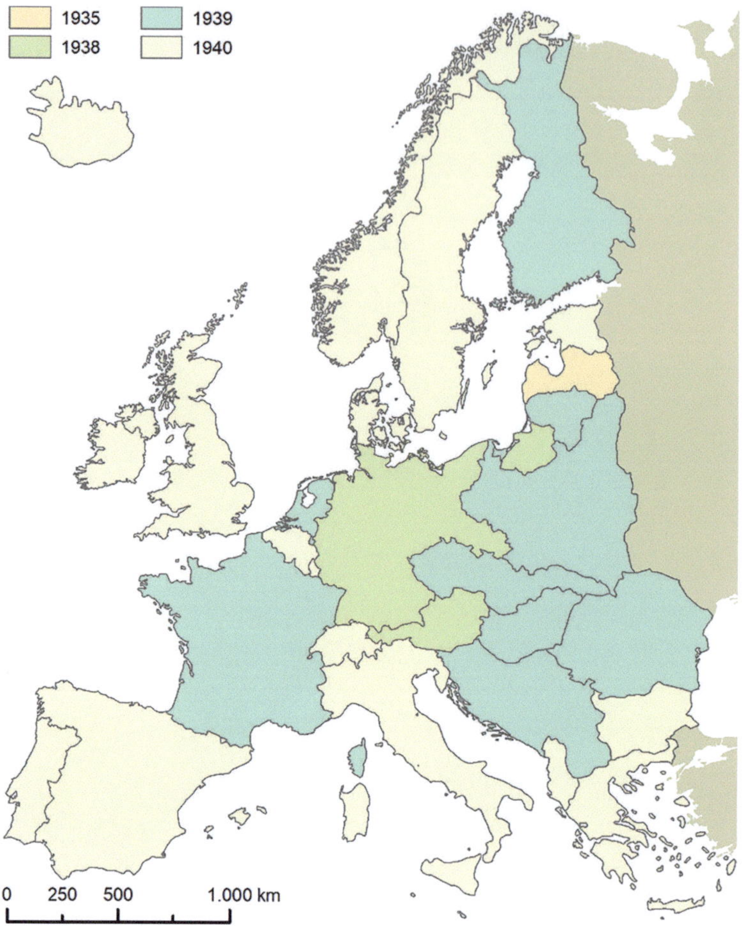

Fig. 3.9 Availability of regional subdivisions in REGIS/REThM, based on the 10-year interval 1940: 1935–40

making it difficult to assign precise borders for 1940 and 1950. The dataset for 1940 reflects the prewar borders of the most affected countries

Figure 3.10: Although the Second World War ended in 1945, its consequences extended well beyond 1950. While most borders were established before that year, Greece regained the Italian Aegean Islands (Dodecanese) in 1947, but administrative information is available from 1951 onwards. Additionally, in 1952, Romania adopted a new constitution that officially transformed the country into a Communist state

Figures 3.11 and 3.12 illustrate and compare all national (L0) and regional (L1 and L2) borders across Europe for the base year 1870 and the end year 2020.

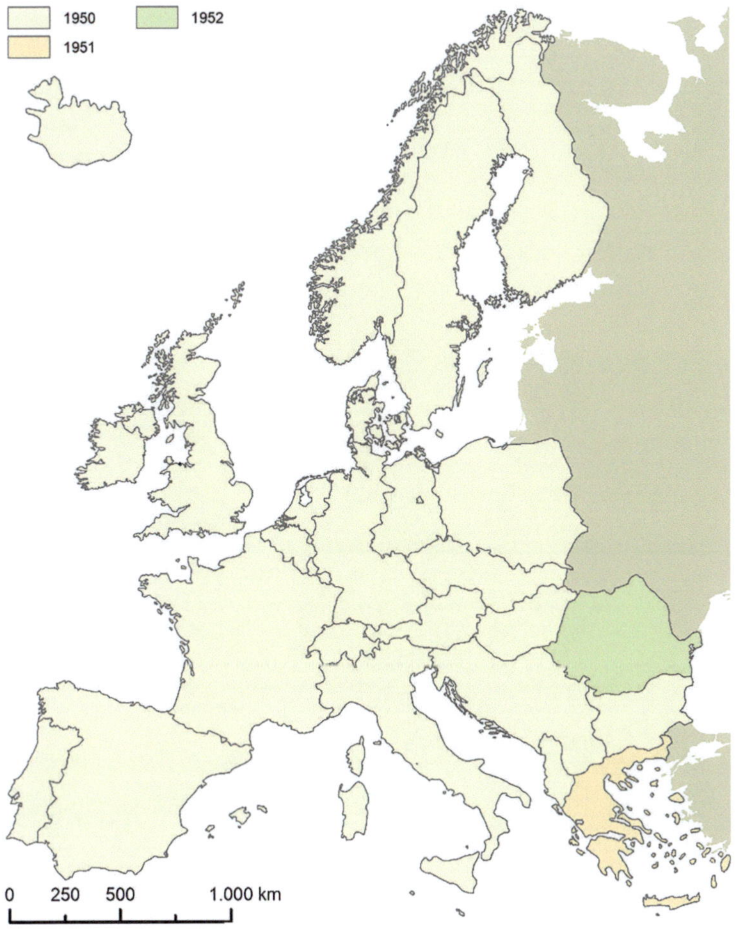

Fig. 3.10 Availability of regional subdivisions in REGIS/REThM, based on the 10-year interval 1950: 1950–52

3.7 Overview of the Download Structure

The GeoPackages containing the historical administrative boundaries of Europe, provided for complete download in one zip file alongside this book/chapter, are organized into three main folders at the top level (see Fig. 3.13).

The main folder, "Sovereign_States," includes subfolders with the GeoPackages for the 47 nation-states represented in the data. Each subfolder contains two GeoPackages corresponding to the two levels of generalization (REGIS and REThM), each of which includes all temporally and administratively differentiated layers for the respective country (see example "REGIS_Czechoslovakia.gpkg" in Fig. 3.13).

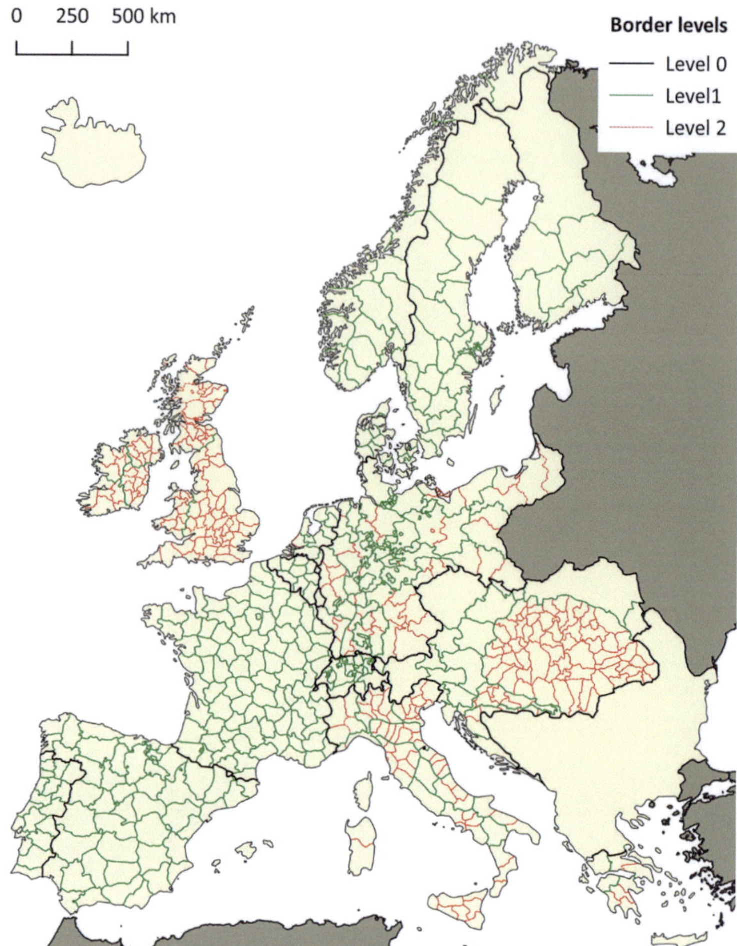

Fig. 3.11 National and regional borders in Europe in 1870 (using REThM)

The temporary semi-sovereign entities that did not receive full international recognition or were subject to the League of Nations (see Table 3.2) are represented in the main folder "Temporary_States_and_Regions" (see Fig. 3.13) and contain the same subfolders as the "Sovereign states" folder.

The main folder, "Pan-European_Coverages," contains the compiled map layers for all of Europe, as described in the previous chapter. It includes, without subfolders, the six GeoPackages for Levels 1–3 of the two generalization levels. Each of these GeoPackages contains a total of 16 compiled layers representing the decades between 1870 and 2020 (see example "REThM_Europe_L2.gpkg" in Fig. 3.13).

When opening a GeoPackage in GIS software such as QGIS, users can choose whether to load all map layers within one GeoPackage or only selected layers.

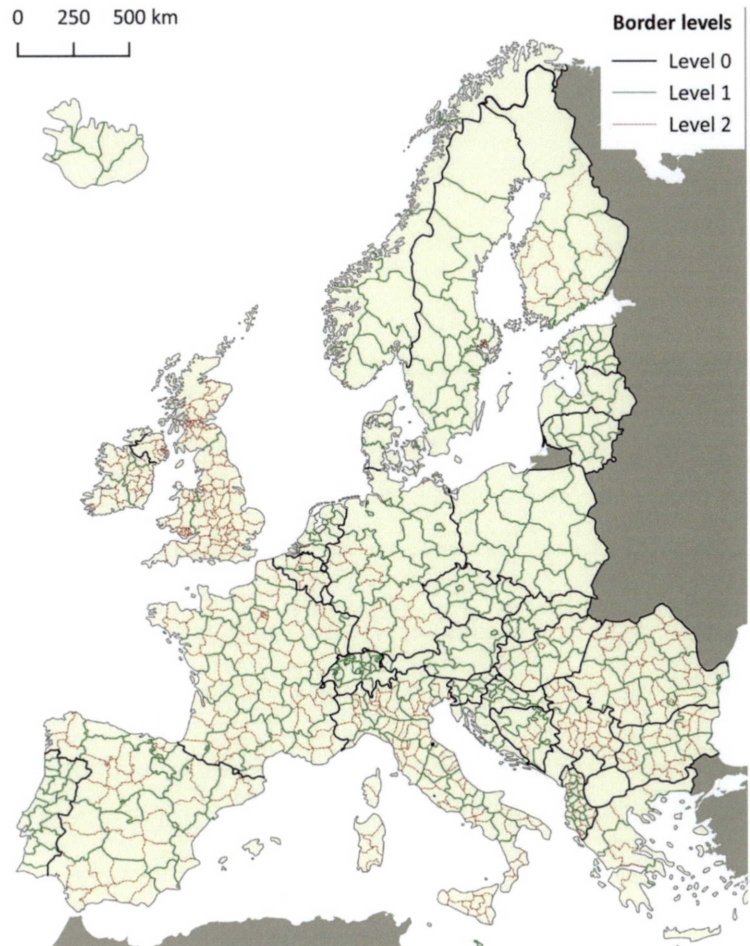

Fig. 3.12 National and regional borders in Europe in 2020 (using REThM)

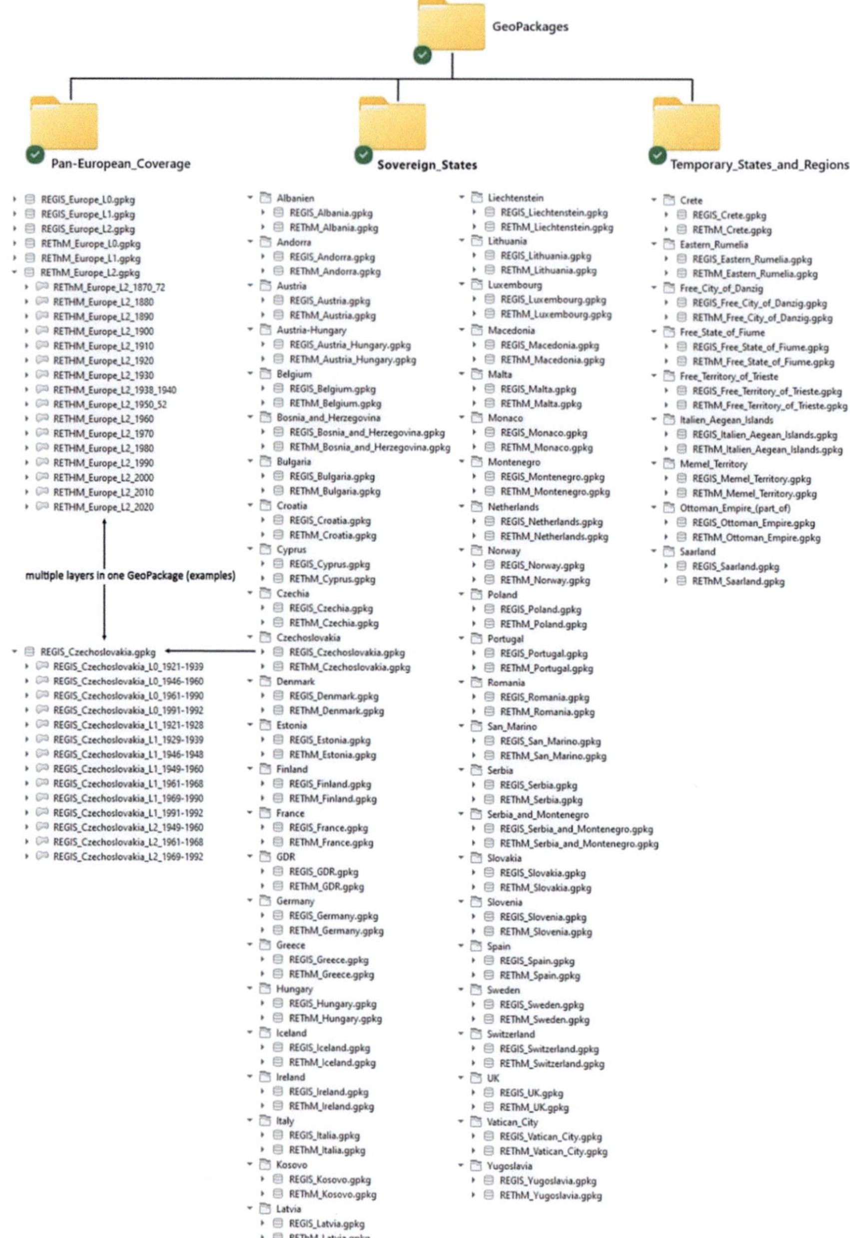

Fig. 3.13 Overview of the file structure within the provided GeoPackages

References

Bettoni, G. (2021). Italy. In Martí-Henneberg, J. (ed.), *European Regions, 1870–2020*. Springer. https://doi.org/10.1007/978-3-030-61537-6_14

Caruana-Galizia, P., & Martí-Henneberg, J. (2013). European regional railways and real income, 1870–1910: A preliminary report. *Scandinavian Economic History Review, 61*(2), 167–196. https://doi.org/10.1080/03585522.2012.756428

European Union. (ed.). (2022). *Statistical regions in the European Union and partner countries. NUTS and statistical regions 2021. 2022 edition.* Publications Office of the European Union.

Eurostat. (2015). *Regions in the European Union. Nomenclature of territorial units for statistics NUTS 2013/EU-28.* Publications Office of the European Union. https://doi.org/10.2785/53780

Gebhardt, H., Glaser, R., & Lentz, S. (Eds.). (2013). *Europa-eine Geographie.* Springer Spektrum.

Gregory, I. N., Martí-Henneberg, J., & Tapiador, F. J. (2009). Modelling long-term Pan-European population change from 1870 to 2000 by using geographical information systems. *Journal of the Royal Statistical Society Series A: Statistics in Society, 173*(1), 31–50. https://doi.org/10.1111/j.1467-985X.2009.00598.x

Martí-Henneberg. (2017). The influence of the railway network on territorial integration within Europe (1870–1950)". *Journal of Transport Geography, 62*, 160–171.

Nagel, K. J. (2021). Germany. In Martí-Henneberg, J. (eds.), *European Regions, 1870–2020.* Springer. https://doi.org/10.1007/978-3-030-61537-6_1

Quick, M. (1994). *Regional territorial units in Western Europe since 1945.* MZES Working Papers, 5, EURODATA, Mannheim.

Triandafyllidou, A., & Gropas, R. (2023). *What is Europe?* Taylor & Francis.

Chapter 4
Use of the REGIS and REThM Data

A variety of GIS software products are available for visualizing and analyzing geospatial data. While proprietary solutions such as ArcGIS provide extensive functionality, open-source alternatives have become increasingly powerful and accessible. Chapter 2 outlined the GIS software products used to generate the REGIS and REThM data. This section examines suitable GIS software options for beginners who wish to perform basic tasks, including displaying regions with administrative boundaries, linking additional attribute data, and creating thematic maps. The focus will be on open-source solutions, though proprietary software will also be briefly discussed.

Among open-source GIS tools, QGIS (QGIS Documentation, 2025) stands out as the most user-friendly and feature-rich solution. It provides an intuitive graphical interface, allowing users to load and visualize geographic data without prior expertise. QGIS supports numerous vector and raster formats, enabling users to import and display country boundaries from sources such as Natural Earth or OpenStreetMap. Additionally, it allows easy integration of attribute data, enabling users to attach and visualize statistical information, such as population density or physician-population ratios. Thematic mapping is straightforward, with built-in tools for categorizing and symbolizing data according to predefined rules (QGIS Documentation, 2025).

Other open-source GIS solutions, such as GRASS GIS and GeoServer, offer extensive geospatial processing capabilities but require a steeper learning curve. GRASS GIS, for instance, is highly powerful for advanced spatial analysis but may be overwhelming for beginners (GRASS GIS Documentation, 2025). GeoServer, on the other hand, is primarily designed for web-based GIS applications rather than desktop use.

ArcGIS by Esri remains a leading commercial GIS solution with comprehensive tools for spatial analysis and cartography. While it is a powerful option, its licensing costs can be a barrier for casual users or those in academia with little previous experience. Beginners may find the interface somewhat complex, but the software offers extensive documentation and tutorials (Esri ArcGIS Documentation, 2025).

© The Author(s) 2026

J. Schweikart et al., *Regional European Geographic Information System (REGIS),*
1870–2020, SpringerBriefs in Geography, https://doi.org/10.1007/978-3-032-14932-9_4

For users with prior programming experience, the open-source programming language "R" (R Manuals, 2025), particularly with the integrated development environment RStudio Desktop and the sf, (Simple Feature for R, 2025), tmap, and ggplot2 packages, provides an alternative approach to GIS. R enables automated spatial data processing, statistical analysis, and thematic mapping, making it a valuable tool for advanced users. However, it lacks an intuitive graphical interface, making it less suitable for complete beginners.

Considering accessibility, functionality, and ease of use, QGIS emerges as the most suitable option for beginners and will be the primary tool used in subsequent sections for demonstrating basic GIS applications. Its combination of an intuitive interface, rich feature set, and strong community support makes it an ideal choice for those new to geospatial data. While other tools such as R can complement GIS workflows, especially for users with programming knowledge, QGIS remains a good and efficient starting point for learning and performing basic GIS tasks efficiently.

4.1 Working with GeoPackage and Thematic Data in QGIS

This section explores the derivation of thematic maps that visualize the evolution of Europe's administrative boundaries. By overlaying boundary configurations from various time periods, we reveal shifts in territorial divisions, offering a clear representation of historical geopolitical transformations.

4.1.1 Opening and Displaying GeoPackage Layers

After downloading, installing, and launching the latest QGIS software (QGIS Download, 2025) and downloading and extracting the REGIS and REThM data (as described at the end of Chap. 3), the layers of the various GeoPackages are ready for visualization in QGIS.

The following section briefly explains two different ways to open layers contained in a GeoPackage. Both methods provide quick and efficient ways to load spatial data into QGIS for visualization and analysis (see Fig. 4.1).

For a better comparison of boundary changes across different time periods, you can adjust the layer styles in QGIS. Simply double-click on the layer name, navigate to Simple Fill, and customize the polygon fill and boundary (stroke) style and color to enhance visual clarity (see Fig. 4.2).

Open GeoPackage Layer Using QGIS Menus

- In QGIS, go to Layer → Add Layer → Add Vector Layer
- Click Browse in the Source field, navigate to the location of the GeoPackage you want to visualize, and select it
- Click Add, and in the next dialog box, choose either to load all layers or select one or more specific layers
- Click Add to load the data into the QGIS map window

Select Items to Add | REGIS_Poland ✕

D:\Dropbox\BHT\REGIS_Projekt\2022_SpringerBriefs_REGIS\GeoPackages\National_States\Poland\REGIS_Poland.gpkg

Search...

Item	Description
REGIS_Poland_L0_1922-1939	MultiPolygon (1)
REGIS_Poland_L0_1946-1950	MultiPolygon (1)
REGIS_Poland_L0_1951-2020	MultiPolygon (1)
REGIS_Poland_L1_1922-1938	MultiPolygon (16)
REGIS_Poland_L1_1939	MultiPolygon (16)
REGIS_Poland_L1_1946-1950	MultiPolygon (14)
REGIS_Poland_L1_1951-1975	MultiPolygon (17)
REGIS_Poland_L1_1976-1998	MultiPolygon (49)
REGIS_Poland_L1_1999-2020	MultiPolygon (16)

Select All   Deselect All

▶ Options

Add Layers   Cancel

Drag and Drop GeoPackage from Windows Explorer

- Open the folder containing your GeoPackage in Windows Explorer.
- Drag and drop the file directly into the QGIS map window
- Click Add, and in the next dialog box, choose either to load all layers or select one or more specific layers
- Click Add to load the data into the QGIS map window

Fig. 4.1 Step-by-step guides to opening GeoPackage layers in QGIS

4.1.2 Linking External Thematic Data

When working with historical administrative boundaries of Europe, researchers may need to integrate external datasets. The geodata offer significant potential for representing external data. Administrative units have long served as the spatial framework for most statistical surveys. Any data collected in Europe over the past 150 years relating to these administrative or statistical units, or which can be aggregated to them, can now be visualized in thematic maps using the geodata produced. Due to

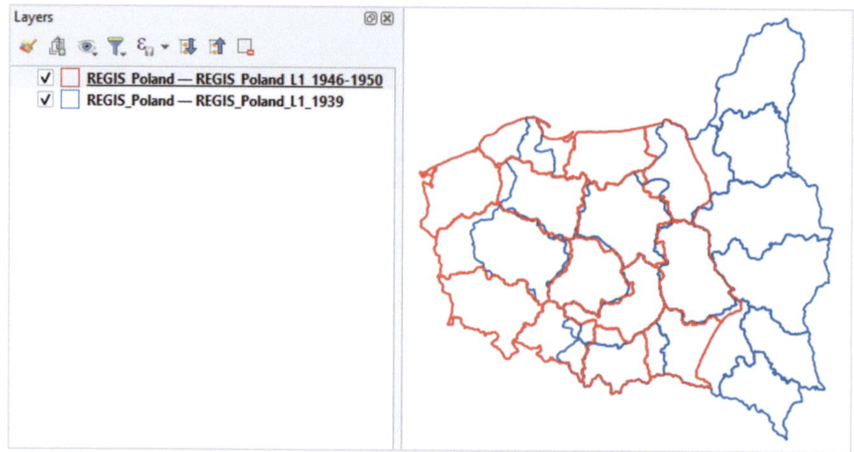

Fig. 4.2   Example of overlaying two GeoPackage layers for Poland in QGIS

the hierarchical structure of the data, all levels can be represented independently of thematic content—whether comparing sovereign states or analyzing the smallest administrative units available in the database within a country. What makes this offering unique is the ability to visually compare data from widely separated time periods in easily interpretable maps, providing a compelling illustration of European societal developments over time.

QGIS allows attribute joins between layers using shared attributes, such as IDs or place names. The ideal scenario is to have a common unique identifier (ID) or consistently spelled names to facilitate accurate linking. However, working with historical data presents unique challenges, because inconsistencies in spelling require additional preprocessing to standardize names.

The first significant challenge in connecting external data to our historical administrative boundaries is that the datasets researchers may wish to integrate are often themselves historical. Many of these sources do not exist in digital form but are preserved in books, archival documents, or handwritten records. To be used in GIS, such materials must first be scanned, processed, and manually digitized. Even when historical data are available online, they are often provided as scanned PDF documents, making direct digital integration impossible. In many cases, researchers need to manually extract relevant information and add it as attributes to the geospatial dataset.

Another significant challenge is that unlike modern classifications such as the NUTS system (Eurostat, 2004), which provides standardized regional IDs, historical administrative units lacked such uniform identifiers. Due to frequent territorial changes, merging or splitting of regions, and evolving naming conventions, applying a single modern classification retroactively is not feasible.

For our historical dataset, it was crucial to develop an internal ID system that ensures consistency within the dataset. This structure allows the representation of parent–child relationships, making it possible to trace which administrative units emerged from others or were merged over time. However, since external datasets will not share this ID structure, linking must rely on matching place names—a process complicated by historical spelling variations, language differences, and inconsistent documentation.

These additional processing steps introduce further complexity, requiring careful handling to ensure historical accuracy and consistency when linking external data to our administrative boundaries.

A common challenge arises from discrepancies in naming conventions, spelling variations, and historical changes in administrative divisions. QGIS provides several tools to facilitate the integration of such datasets effectively. Tools such as OpenRefine or Python libraries like pandas and fuzzywuzzy can help clean and normalize data before importing it into QGIS. The fuzzy matching approach, available through Python plugins in QGIS or external tools like R, enables approximate string matching to link datasets with minor spelling differences.

Using a simple example in which historical data are manually linked to a layer in QGIS, we aim to demonstrate the potential of these data for historical analysis. The Central Statistics Office (2025) of Ireland provides access to digitized historical census reports for download. These reports have been scanned and are available as PDFs, rather than in structured digital table formats. For example, Volume 1 of the 1926 Census of Population includes a PDF titled "Increase or decrease percent in population of Saorstát Éireann and of each Province or County as constituted at each of the eleven Censuses from 1821 to 1926."

As seen in Fig. 4.3, the PDF provides the percentage increase or decrease in population for Irish counties over different census periods, including the period from 1911 to 1926. In our geodata collection, the "Sovereign_States" folder contains a subfolder named "Ireland," where the GeoPackage "REGIS_Ireland.gpkg" can be found. When opened in QGIS, as described in the previous section, the various layers of the GeoPackage are displayed. Counties are represented at administrative level 2 (L2), making the "REGIS_Ireland_L2_1922-1929" layer the appropriate choice for visualizing population changes between 1911 and 1926. Administrative units for Ireland before Ireland's independence in 1922 are stored within the UK folder, reflecting its historical status as part of the United Kingdom during that period.

To manually add the data from the PDF to the GeoPackage layer in QGIS, follow the step-by-step instructions in Figs. 4.4 and 4.5.

When manually entering historical statistics, a common issue arises: the old census data lists "Tipperary County," but this entity does not exist in the geodata. Instead, the dataset contains two separate counties, Tipperary North Riding and Tipperary South Riding, which are not individually mentioned in the statistical report.

Historically, Tipperary was a single county when census records were first introduced, and it continued to be reported as such in statistical yearbooks. However, by 1922, Tipperary had already been divided into two separate administrative counties, which is accurately reflected in the geodata. Interestingly, in modern times, these two counties have been reunited into a single entity.

There are two possible ways to address this discrepancy in QGIS:

County or Province.	Percentage Change ($+$ = Increase) ($-$ = Decrease)					
	1861 to 1871	1871 to 1881	1881 to 1891	1891 to 1901	1901 to 1911	1911 to 1926
LEINSTER :						
Carlow County	$-$ 9.60	$-$ 9.84	$-$12.09	$-$ 7.79	$-$3.96	$-$ 4.90
Dublin City and County	$-$ 1.22	$+$ 3.37	$+$ 0.07	$+$ 6.92	$+$6.47	$+$ 5.96
Kildare County... ...	$-$ 8.06	$-$ 9.34	$-$ 7.38	$-$ 9.46	$+$4.82	$-$12.91
Kilkenny County ...	$-$12.16	$-$ 9.00	$-$12.33	$-$ 9.28	$-$5.30	$-$ 5.30
Laoighise County	$-$12.00	$-$ 8.33	$-$11.27	$-$11.51	$-$4.86	$-$ 5.65

Fig. 4.3   Excerpt from a historical census document, Ireland, 1926 (Central Statistics Office, 2025)

Open the Attribute Table

- Ensure the REGIS_Ireland_L2_1922-1929 layer is loaded in QGIS (as described in the previous section).
- In the Layers panel (usually on the left side of the screen), right-click on the layer name.
- Select "Open Attribute Table" from the menu. This will display a table containing all attributes (data fields) associated with the counties.
- Locate the column labeled "name," which contains the county names.

    ...

Create a New Column

- In the Attribute Table window, look for the toolbar at the top and click on the "Toggle Editing Mode" button (a small pencil icon). This allows modifications to the data.
- Click on the "New Field" button (an icon with a table and a small plus sign).
- A dialog box will appear where you can define the new field:
- Name: "1911 to 1926" (to indicate the population change period).
- Type: "Decimal number (real)" to allow decimal values.
- Click OK to create the new column.

Fig. 4.4   Step-by-step instructions on adding data from the PDF to a GeoPackage layer (Part 1)

- Duplicate the Population Change Value—The percentage change for Tipperary County from the census data can be entered into both North Riding and South Riding, ensuring that both regions retain the historical information.
- Merge the Two Polygons—Alternatively, the two Tipperary polygons can be merged into a single entity in QGIS using the "Dissolve" tool. This approach

---

Manually Enter the Population Change Values

- Now, for each county, manually match the names from the PDF to the "name" column in QGIS.
- In the newly created "1911 to 1926" column, enter the corresponding percentage increase or decrease in population as found in the PDF.

Save the Changes

- Once all values have been entered, click on the "Toggle Editing Mode" button again to exit editing mode.
- QGIS will ask if you want to save the changes – confirm by clicking "Save".

Fig. 4.5 Step-by-step instructions on adding data from the PDF to a GeoPackage layer (Part 2)

aligns the geodata with the statistical records but may impact spatial analyses that require the original subdivisions.

Choosing the appropriate method depends on the specific research goals and whether maintaining the administrative division of 1922 or aligning with the historical census format is the priority.

Now, the REGIS layer contains the historical population change data for 1911–1926, making it available for further analysis and visualization in QGIS. In the next chapter, we will create a thematic map of the historical statistics using that layer.

4.1.3 Creating Thematic Maps from Linked Data

In the previous chapter, historical census data on population change between 1911 and 1926 were manually linked to the QGIS layer. The following section provides a step-by-step guide for QGIS beginners to create a thematic choropleth map using this new attribute.

To ensure that both population increase and decrease are clearly distinguishable, we will use a diverging color scheme. However, since only one region experienced population growth while all others declined, we will apply the following approach:

- One class for population increase, using a distinct color.
- Multiple classes for population decline, with varying shades to reflect different levels of decrease.

This method ensures that even small changes in population decline are visually represented while maintaining clear differentiation from the single instance of population growth. The following step-by-step instructions will guide you through styling the map in QGIS. To style the map in QGIS, follow the step-by-step instructions in Fig. 4.6 and the settings in the Layer Styling dialog box in Fig. 4.7.

Open the Layer Styling Panel

- Select the REGIS_Ireland_L2_1922-1929 layer in the Layers panel.
- Click on the "Open the Layer Styling panel" button (paintbrush icon) or go to Right-click → Properties → Symbology.
- In the Symbology section, change the "Single Symbol" option to "Graduated".

Configure the Color Scheme

- In the Value dropdown, select "1911 to 1926" (the attribute containing population change).
- Click on the Color ramp and select a diverging color scheme (e.g., under "All color Ramps" red to blue (RdBu), where red represents population decline and blue represents growth).
- Click on Classify to automatically generate value ranges.

Adjust the Classification (Since there is only one region with a population increase, we need to manually adjust the classification)

- Set a single class for positive values (e.g., 0 to max value) with a blue color.
- For negative values, create multiple classes and assign them a gradient of red shades to distinguish varying levels of decline (see example, Fig. 4.4).
- Click Apply and OK to save the settings.

Fig. 4.6 Step-by-step instructions on styling the map in QGIS

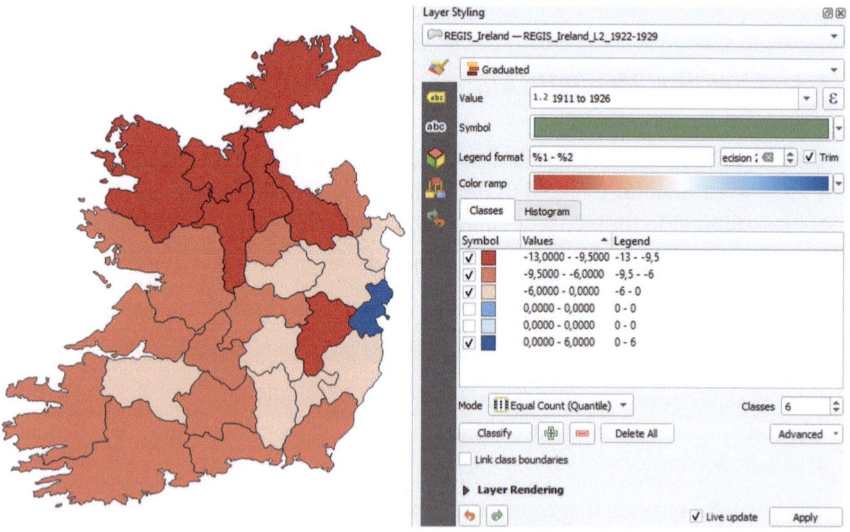

Fig. 4.7 Map with Layer Styling dialog

To enhance the readability of the map, we can add county labels displaying their names. To do this, open the Layer Styling panel, switch to the Labels tab, and select Single Labels. In the Value dropdown, choose the "name" attribute to label each county accordingly. Adjust the font size, color, and placement to ensure clarity and avoid overlapping text.

Once the map styling is complete, we can create a final map layout for presentation. Open the Print Layout via Project → New Print Layout, give it a name, and confirm. In the layout editor, click Add Map, then draw a rectangle to place the map. Add a title using the Add Label tool and position it appropriately. To ensure proper interpretation, insert a legend via Add Legend, adjusting its position and formatting as needed. Do the same for a map scale via Add Scale Bar. Finally, use the Export function to save the completed map as an image or PDF for further use.

Figure 4.8 shows an example of a fully developed thematic map illustrating the spatial patterns for interpreting population change in the Irish counties between 1911 and 1926. The significant population decline in Ireland between 1911 and 1926 was driven by war, political instability, economic hardship, and continued emigration. The effects of the World War I, the Irish War of Independence, and the Irish Civil War led to widespread insecurity and economic decline, particularly in rural areas. Many people emigrated in search of better opportunities, especially to Britain, the United States, and Canada.

Dublin was the only county to experience population growth during this period. As the new capital and economic hub, it offered better employment opportunities, attracted administrative and commercial activity, and became a refuge for people leaving struggling rural areas. While the rest of Ireland faced severe demographic losses, Dublin's relative stability and economic importance made it an exception to the overall trend.

The more significant population decline in the northern counties can be attributed to a combination of economic factors (lower levels of industrialization and development opportunities), political and social unrest (Irish War of Independence), as well as migration (especially to British Northern Ireland). In contrast, the southern counties, particularly around Dublin, were more influenced by urban development, better economic conditions, and more stable political circumstances, which contributed to a comparatively smaller population decline.

4.2 Use Cases and Application Examples

The following chapter presents a series of application examples demonstrating the potential of our historical administrative boundary dataset for research and teaching. Through four case studies, we illustrate how these geospatial data can be enriched with additional information to generate thematic maps that provide valuable insights into historical developments across different regions and time periods.

In the first chapter of this section, we focus on Scandinavian population trends. By integrating census data from 1900 and 1970 into the geospatial dataset, we generate

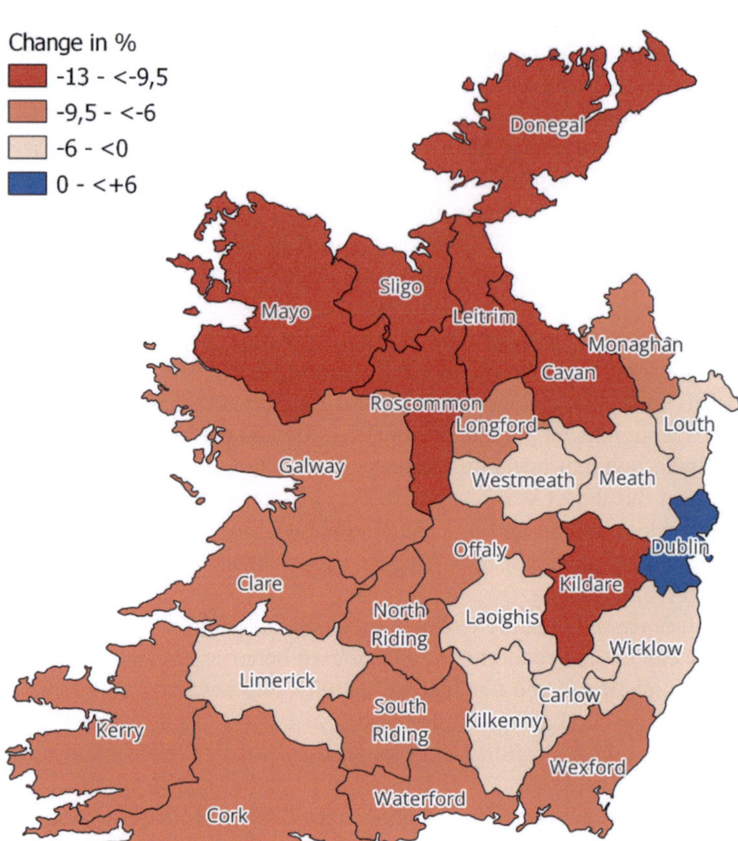

Fig. 4.8 Example of a fully developed thematic map with essential elements

thematic maps of population density at these points in time. The comparison of these maps illustrates demographic changes in the region and highlights long-term patterns in population distribution.

In the second section, the map collection provides a view of historical and administrative changes, offering deeper insights into European territorial history. These maps, created through graphic design software rather than GIS, combine high-quality cartography with visual clarity and historical context, supporting further research and learning.

The third section examines infant mortality rates across Europe around 1990. By enriching the dataset with historical statistics, we create a thematic map that enables a comparative analysis of infant mortality across administrative regions at that time.

This visualization provides a unique perspective on regional disparities in public health and living conditions at the time.

The final section presents an analysis of the historical development of physician density in Germany from 1876 to 2000. Over this period, not only did Germany's external borders change, but its administrative structures were also redefined multiple times, making direct comparisons between early and modern regions challenging. To address this issue, we apply a grid-based method for change detection, which allows historical data to be harmonized with modern administrative boundaries. This approach facilitates meaningful comparisons of healthcare provision over time.

Together, these case studies illustrate the diverse ways in which our geospatial dataset can be applied to a range of disciplinary research questions. Whether for historical geography, demography, public health, or political science, the ability to integrate and visualize historical statistical data in a geospatial framework opens new avenues for analysis and interpretation.

4.2.1 Population Change in Scandinavia

Marti-Henneberg (2005) demonstrates through an analysis based on historical population data for European administrative regions that the most densely populated areas of Europe have experienced continuous population growth since 1870, further widening regional population disparities. We aim to test this hypothesis using our geospatial data with a selected empirical case study.

Scandinavia is particularly well-suited to test this hypothesis due to its long-term political stability and well-documented historical population data, which enable consistent regional comparisons over time. Additionally, the region exhibits a pronounced south-north gradient in population density, with densely populated urban centers such as Stockholm, Copenhagen, and Oslo in the south, while the northern regions remain sparsely inhabited. This strong contrast makes it an ideal case for analyzing whether population growth has been concentrated in already densely populated areas, further widening regional disparities over time.

Examining population density in the census years 1900 and 1970 is particularly useful because these dates capture key phases of demographic and socioeconomic change in Scandinavia. In around 1900, the region was still largely agrarian, with industrialization and urbanization just beginning to accelerate, providing a snapshot of pre-industrial population distribution. By 1970, Scandinavia had undergone extensive urbanization, economic transformation, and welfare state expansion, significantly altering settlement patterns (Schön, 2007). Comparing these two points in time allows us to analyze whether population growth reinforced existing disparities, with urban centers expanding while rural areas experienced stagnation or decline.

In Norway, Sweden, and Finland, a population census was conducted in 1900, while Denmark held its census in 1901. In 1970, censuses were carried out in all four Scandinavian countries. After researching the population data for these census years at the administrative regional level, the appropriate layers from the GeoPackages of our geodata are selected and supplemented with the population, using the method outlined in this chapter.

To represent the entire Scandinavian region on a single map sheet, it is advisable to use the REThM data, as the numerous islands and fjords along the Scandinavian coast have been generalized (simplified). Detailed coastlines like within the REGIS data at small scales would clutter the map and reduce readability. The population is then related to the area of each region to calculate population density. A choropleth map displaying absolute population is not appropriate.

To enhance comparability, a consistent color gradient is designed for both census years. Figure 4.9 presents the thematic maps for the two census years side by side.

The comparison of the two thematic maps reveals a clear south-north gradient in population density as early as 1900/1901. The sparsely populated areas are concentrated in the northern parts of Norway, Sweden, and Finland, while Denmark and the southern regions of these countries exhibit significantly higher population densities. The 1970 map highlights how this pattern has intensified over time. The already densely populated southern regions have experienced further growth, with urban centers and their surrounding areas standing out. In contrast, population density in the northern regions has remained largely unchanged. This example supports the

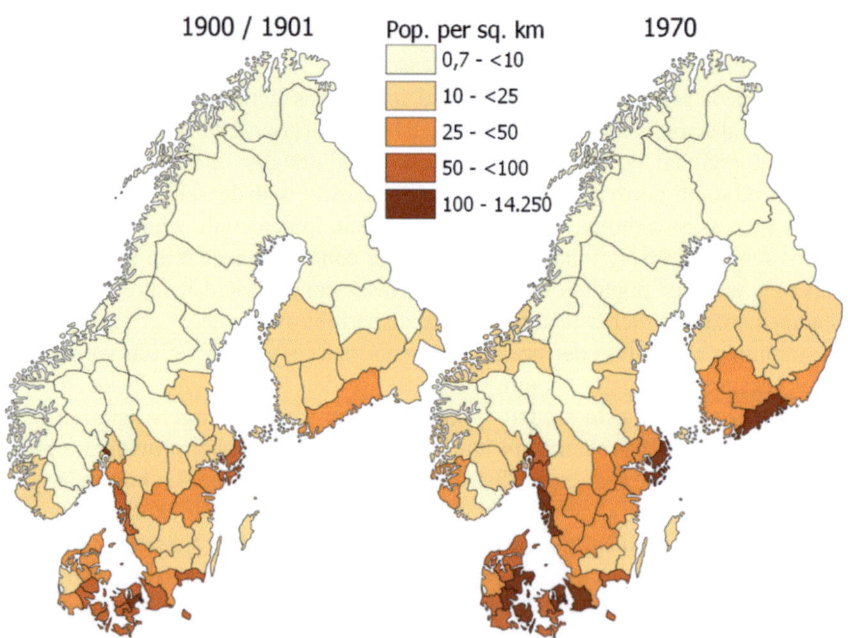

Fig. 4.9 Map of population growth in Scandinavia

hypothesis introduced at the beginning of the chapter, demonstrating how population disparities have increased over time in Scandinavia.

4.2.2 Deriving a Collection of Historical Thematic Maps

By integrating maps into the analysis of historical and administrative changes, the collection offers a multidimensional perspective that text alone cannot provide. The maps serve as a critical tool for researchers, educators, and students, enabling a deeper understanding of the complexities of European territorial history. They also pave the way for further research by providing a visual framework upon which future studies can build.

This following map collection showcases visualizations derived from our historical geodata but not produced within a GIS environment. Instead, the data were post-processed using professional graphic design software. The maps represent a condensed and stylized output of a high-quality cartographic production process that prioritizes visual clarity and historical interpretation. While based on accurate geospatial sources, the workflow behind these maps involved design decisions and manual refinements that go beyond standard GIS procedures and may not be easily replicated by non-specialists.

The Map Collection

This section presents the "Map Collection European Regions, 1870–2010," a comprehensive series of maps derived from the historical geodatabase. These maps offer a detailed and structured visual representation of administrative boundary changes across Europe, enhancing the analysis of geopolitical transformations over time. By systematically compiling, refining, and standardizing historical data, this collection becomes an indispensable resource for understanding territorial evolution, regional development, and the historical trajectory of state formation and administrative restructuring. The collection includes 104 thematic maps (see Fig. 4.10), which were originally created as illustrations for a publication derived from the REGIS Project (Martí-Henneberg, 2021). Although the data come from the REGIS project, the maps were developed in direct alignment to conceptual framework and content of the book. They were not simply by-products of the original project, but rather carefully designed elements intended to complement and enrich the publication. These maps serve as a bridge between textual analysis and visual representation, allowing readers to grasp complex territorial and administrative changes through clear, consistent, and visually engaging graphics.

At the core of this project are the territorial units and, consequently, the territorial history of Europe. No medium is better suited to visualizing these aspects than maps. Therefore, the creation of maps is a logical and essential step in this endeavor. In the volume edited by Martí-Henneberg (2021), national and regional border changes in Europe between 1870 and 2020 are analyzed. In addition to the introductory chapters, the book includes 27 country-specific chapters that enable an understanding

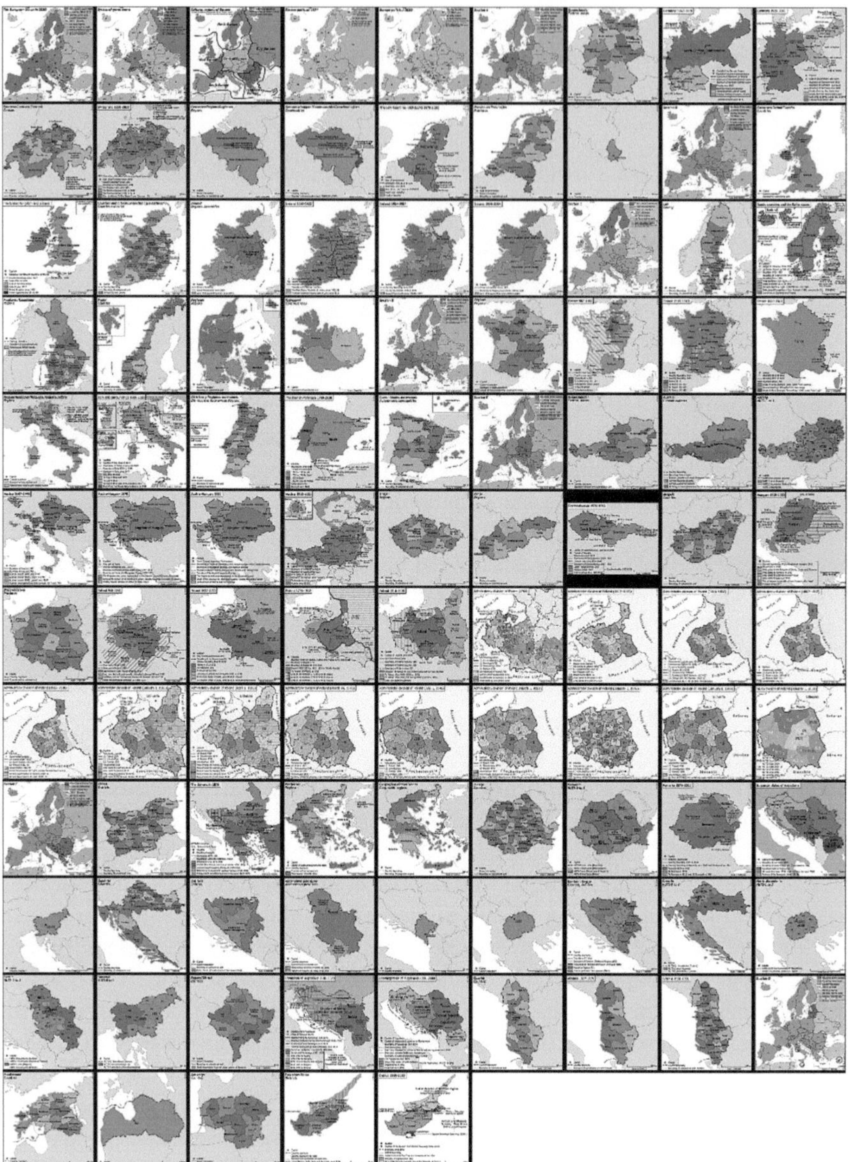

Fig. 4.10 All maps in order of their appearance in the publication (Marti-Henneberg, 2021)

of border changes at various levels of territorial organization. These texts describe the process of state formation in relation to regional-administrative structures. Each chapter contains maps derived from the database documented in this book.

Maps can serve a variety of purposes to enhance the understanding of historical, administrative, and political developments. Five key aspects are particularly relevant:

- **Spatial Contextualization**: They anchor historical and administrative discussions within a clear geographic framework, making it easier to understand where events unfolded.
- **Visualization of Administrative Boundaries**: By showing boundaries at different levels, maps highlight the organizational complexity of various regions.
- **Tracing Large-Scale Border Changes**: Maps provide a visual timeline of political and territorial shifts, offering insights into key historical turning points.
- **Thematic Representations of Broader Trends**: Beyond borders, maps reveal patterns of political, cultural, and economic developments across regions.
- **Comparative Analysis**: Side-by-side map comparisons enable a deeper understanding of how different regions or states underwent territorial and administrative changes over time.

There are three map categories in the collection (see Fig. 4.11). "Country maps" focus on current political-administrative structures. They highlight the first level of political-administrative units according to the statistical nomenclature of Nomenclature of Units for Territorial Statistics (NUTS), making them the opening maps for each country chapter. This category also included additional maps on other current lower NUTS level structures and topics.

"Historical maps" document boundary changes and historical developments over time. These include both maps that illustrate changes across broader periods and those that highlight specific historical condition of the administrative units.

"Europe maps" as a smaller subset addresses topics of continental relevance and serves as overviews for specific chapters. While derived from the historical map category, these maps provide higher-level summaries that tie individual chapters into a cohesive narrative.

In many cases, multiple maps are created for a publication. This presents challenges, as the diverse methodological approaches employed by various authors must be harmonized, and maps often need to accommodate different scales. A coherent and consistent layout across all contributions is essential to support the reader's understanding, ensuring clarity and cohesion throughout the publication.

Martí-Henneberg's work (2021) included several individual country chapters, each written by experts. The development of the map collection in addition to the text was iterative and time-intensive. Early efforts focused on creating European maps for the introductory chapters, followed by the gradual addition of country-specific maps drawn from the REGIS database. Additional thematic maps were developed in collaboration with chapter authors, often requiring supplementary data and meticulous adjustments to ensure clarity and consistency. Creating multiple maps for a single publication involves reconciling different methodological approaches, aligning scales, and maintaining a coherent visual style. Ensuring that all maps shared a unified layout and design language was essential for readers to navigate the publication with ease. As a result, every map underwent multiple rounds of refinement. This iterative process involved both technical adjustments—such as correcting projections and standardizing fonts—and aesthetic decisions to ensure that the maps not only

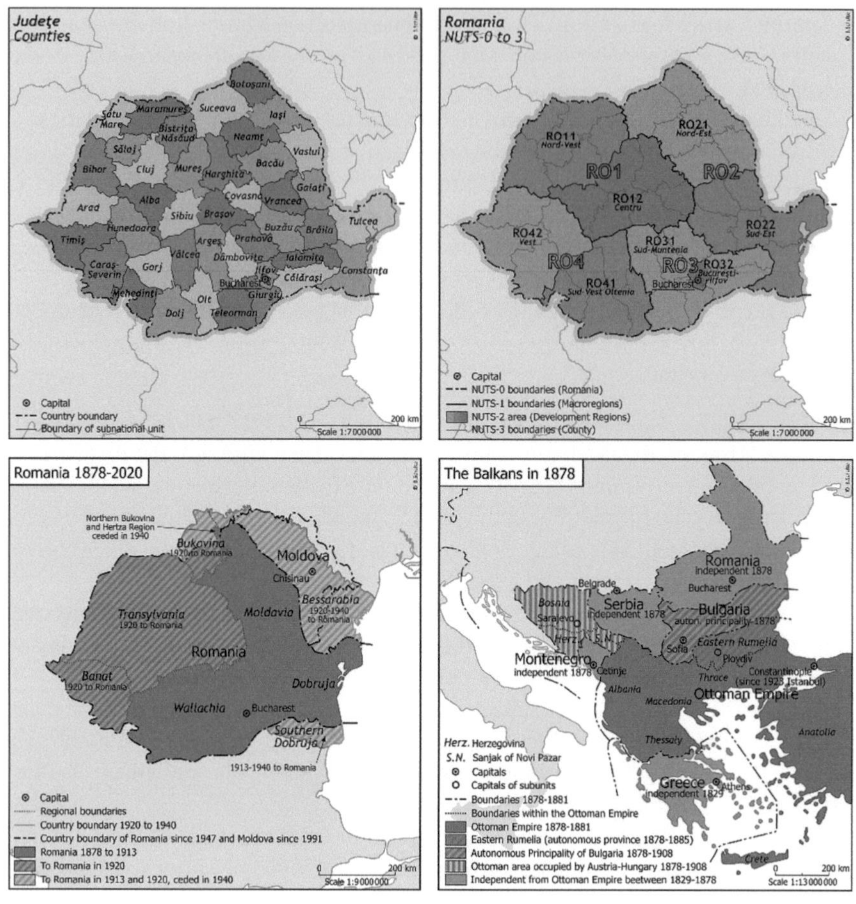

Fig. 4.11 Country maps of Romania (first row) and historical maps (second row)

conveyed accurate information but were also visually compelling and easy to read in printed and digital distribution form.

General Framework for the Map Collection

Before establishing the cartographic concept, it was essential to define the overarching framework that would guide the development of the maps. Several key parameters were considered, including the intended format of the maps, their physical and technical requirements, the choice of projection, color schemes, and the selection of map elements.

First, to standardize the visual layout and ensure consistency across maps of varying geographic scope—from continental overviews to detailed depictions of smaller regions—a fixed map size of 12 × 12 cm was established, and the map scale was subsequently adjusted to the specific area being represented.

Second, to enhance usability, readability, and clarity, map elements were deliberately simplified to include only essential components. These consisted of a clearly stated title and subtitle and legends, ensuring consistency in placement and appearance. These consisted of a clearly stated title and subtitle, names of administrative units and their capital cities, a graphical and numerical scale, and a simplified legend. Superfluous elements such as grid lines, coordinate systems, and projection references were omitted to avoid visual clutter and to keep the focus on the administrative and territorial information. Inset maps were generally avoided but included where necessary, such as for remote territories or islands. Furthermore, highly legible and aesthetically suitable fonts, specifically Tahoma and Trebuchet MS, were selected to support clear interpretation and overall visual coherence. To ensure a consistent and clear layout, the following placement guidelines were applied:

- **Title Placement**: Upper-left corner, clearly indicating the map's subject.
- **Legend Placement**: Bottom-left corner, containing only necessary symbols and descriptions.
- **Scale Representation**: A combined graphical and numerical scale was positioned in the bottom-right corner for consistency.

Third, to ensure visual clarity in black-and-white and grayscale formats, a limited grayscale palette was adopted. Rather than relying on intricate patterns or high-contrast hatching, the maps utilized subtle gradations of gray tones. These tones were extensively tested to guarantee sufficient contrast and readability, even in standard black-and-white print settings. Table 4.1 summarizes the assignment of the gray tones.

Fourth and final guideline, the selection of an equal-area projection ensured accurate representation of territorial extents. Given the focus on administrative units and their spatial development, the Lambert Azimuthal Equal Area projection was chosen. This projection preserves area relationships, making it particularly suitable for visualizing changes in administrative boundaries over time. The projection's center was

Table 4.1 Assignment of the gray levels to the objects displayed on the map

Proportion of black	Map objects displayed
Black (100% black)	Main information, for country borders, main labels and capital signatures for country under review
Darker tone (80% black)	For lines for subnational boundaries and coastline of country under review
Intermediate grays (40%, 30%, 20%, 10% black)	For fills to distinguish different subnational administrative units
Light gray tones (5%, 10% black)	For subtle background differentiation without overpowering the map's primary features like land areas outside the country under review
White (0% black)	Water surfaces

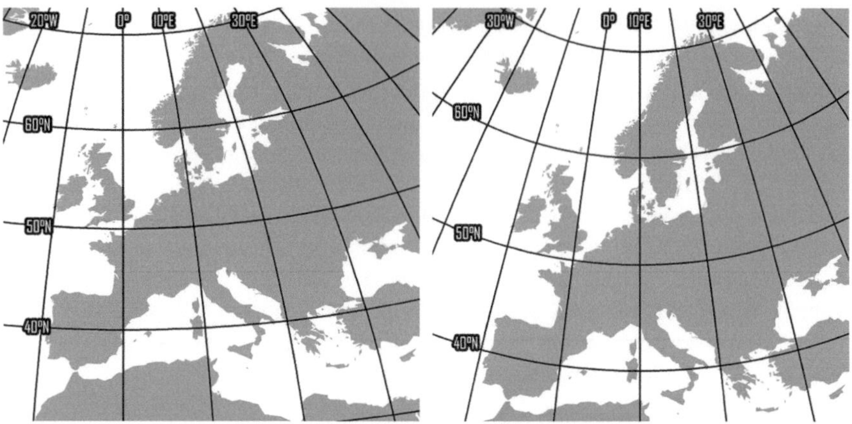

Fig. 4.12 Modified Lambert Azimuthal Equal Area projection with center 50°0′ N, 10°0′ E (right)

set at 10°E and 50°N, near Arnstein in Bavaria, Germany, to minimize distortion across Europe and provide a consistent spatial reference for all maps (see Fig. 4.12).

A balance between automation and manual adjustments was maintained to ensure both efficiency and accuracy in the map creation process. Although much of the cartographic process was standardized, fine-tuning remained critical. Labels were adjusted manually to avoid overlaps, and boundaries were refined to maintain consistency in appearance. This step-by-step refinement was guided by the Pareto principle, recognizing that the final 20% of quality enhancement required a disproportionate effort—yet was crucial for producing a professional result.

Example Production Process for Romania

A vector graphics program was used to implement the project. With the aid of the additional software Avenza MAPublisher, georeferenced data could be processed. The production process began by establishing a systematic layer structure within the graphic software.

Data import and styling began with importing geographic base data, such as REGIS_Europe_L0.gpkg, into a pre-prepared country map template, including land, water, and neighboring countries. The map scale was set to 1:7,000,000 to fit the 12 × 12 cm format. Using the MAPublisher add-on, the required geometry and attributes were imported and styled according to predefined standards. For Romania, the first-level political-administrative units, "județe" (counties), were imported as NUTS-3 units. Administrative boundaries were styled using a grayscale scheme (e.g., 10%, 20%, 30%, 40% black) to ensure clear visual differentiation. Labels for administrative units were applied through MAPublisher's labeling feature and manually adjusted for proper placement, especially in crowded areas. A predefined point signature was manually placed in Bucharest, including a name label.

Cartographic refinements included adjusting country boundaries in coastal areas and manually adding subnational unit water boundaries. Labels and features were

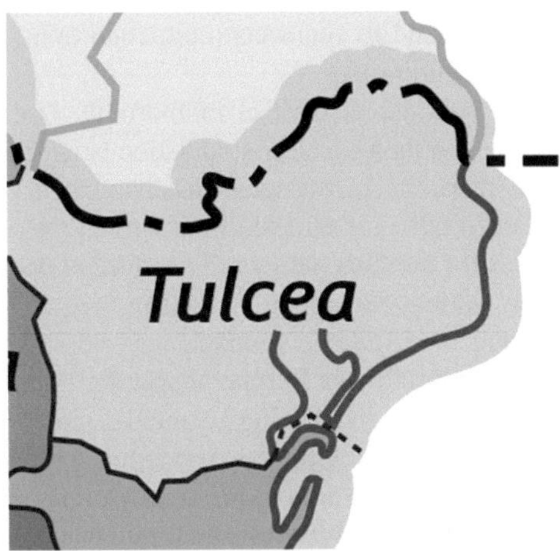

Fig. 4.13 Cartographic perfection of details

repositioned to avoid overlaps, using dedicated layers to ensure legibility in cases of label conflicts with darker features (see Fig. 4.11, top left).

After the map had reached a satisfactory level of refinement, final quality assurance was carried out, including a consistency review. The visual hierarchy was checked to ensure that key administrative units stood out and neighboring country boundaries were de-emphasized. The grayscale fills were confirmed to be distinguishable, and the legend and scale bar were verified for accuracy. With all layers properly aligned and styled, the map was prepared for export in the required format, ready for inclusion in the final publication (see Fig. 4.13).

4.2.3 *Infant Mortality in Europe*

The infant mortality rate (IMR) is defined as the number of deaths of children under one year of age per 1,000 live births in the same year (Reidpath & Allotey, 2003). IMR is not only a key indicator of child health but also serves as an explanatory variable reflecting the socioeconomic development of a country (Gonzalez & Gilleskie, 2017). The specific impact of healthcare systems on population health and, consequently, on infant mortality remains a complex issue (Wolfe, 1986).

Europe has historically exhibited the lowest infant mortality rates globally. Among EU countries, Romania and Bulgaria consistently recorded the highest IMR levels throughout the period from 1994 to 2015, while the Scandinavian countries (Finland, Sweden) maintained the lowest rates (Onambele et al., 2019). This development is part of a long-term historical process. One consistent observation across all time periods is that infant mortality displays substantial regional disparities. Even in the early 1990s, while Western and Northern European countries had already achieved very low mortality rates, parts of Southern and Eastern Europe still exhibited comparatively high levels. These differences not only reflect the quality of medical care, but are also linked to socioeconomic conditions, education levels, infrastructure, and political frameworks.

The regional disparities in mortality rates are complex. However, numerous studies have shown a clear association between socioeconomic inequality and higher infant mortality in the most disadvantaged groups. These findings underline the continued need for targeted interventions to improve maternal health and increase support for families with young children (Doetsch et al., 2023; Harpur et al., 2021; Zilidis & Hadjichristodoulou, 2020).

Using the available geodata, it is possible to visualize the spatial distribution of infant mortality in Europe at specific points in time. Such maps may serve as a starting point for identifying potential causes of observed disparities. GIS technology further allows the temporal development of infant mortality rates to be analyzed by converting data across different administrative boundaries. However, the availability and comparability of data for the regions under investigation must be carefully assessed.

Current data on infant mortality are available from Eurostat. Comparable figures are provided at the NUTS-2 level from 1990 onwards, both as absolute numbers (Eurostat, 2025a) and as rates (Eurostat, 2025b). However, data coverage—especially for the 1990s—remains incomplete. In such cases, statistical sources such as national yearbooks must be consulted. Working with historical data requires matching the regional units found in these sources with the regional codes of the database, which can be time-consuming in individual cases.

For the year 1991, data on infant mortality across Europe were compiled for the whole of Europe from a variety of sources, always using the smallest available regional level. These data were then linked to the corresponding layers from the REThM dataset. The data were classified using quantiles and are presented as a choropleth map in Fig. 4.14. The map clearly illustrates the substantial spatial disparities across Europe. Significant differences can also be observed within individual countries, such as the distinct north–south divide in Italy. Similarly, the countries of Eastern Europe exhibit markedly higher levels of infant mortality.

According to Eurostat's "Statistics Explained," one of the drivers behind the increase in life expectancy at birth in the EU has been the sharp decline in infant mortality rates. As of 2023, the infant mortality rate in the EU stands at 3.3 deaths per 1,000 live births. The most substantial reductions were generally observed in countries that had above-average rates in earlier years. Overall, this decline has contributed to the rise in life expectancy over the same period (Eurostat, 2025c). Nevertheless, current maps still reflect some of the spatial structures observed in 1991, with higher infant mortality persisting in the eastern parts of the EU.

4.2.4   Change Detection: Physician Density in Germany

Germany has undergone significant changes in both its external and internal administrative boundaries over the past 150 years. The administrative divisions of the German

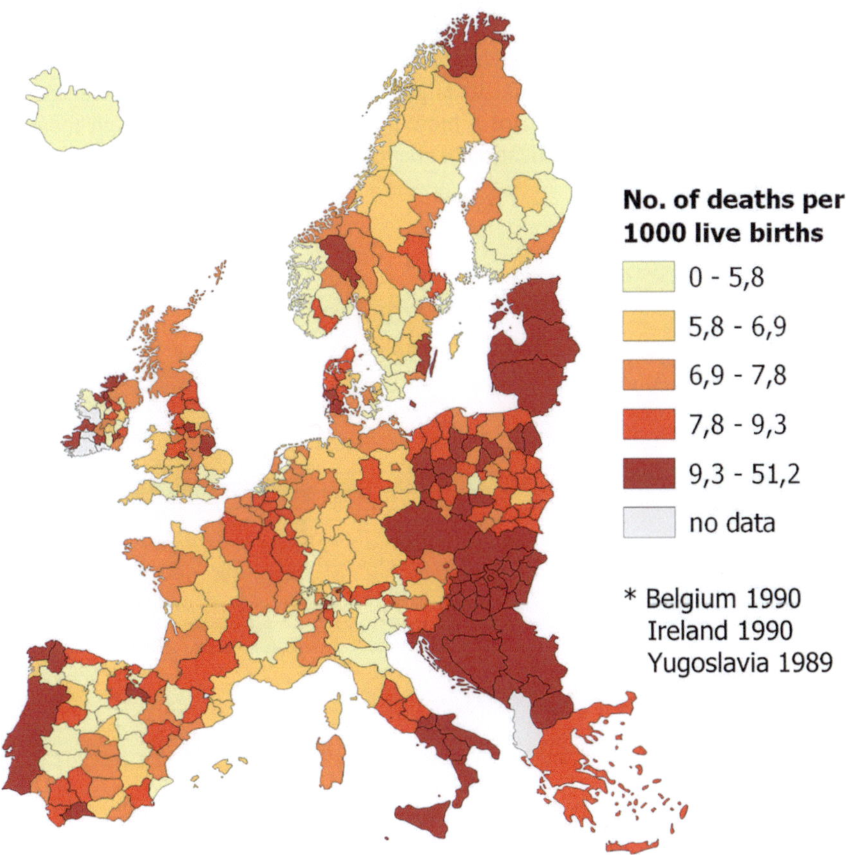

Fig. 4.14 Map of infant mortality rate in Europe, 1991

Empire before World War I bear little resemblance to those of modern Germany, making direct comparisons of historical and contemporary data challenging. To overcome this issue and enhance comparability, we apply a methodological approach that integrates historical physician density statistics with our geodata. This involves converting vector-based historical boundaries into raster data, incorporating physician density as an attribute. By overlaying these rasterized datasets with current administrative boundaries and applying zonal statistics, historical physician density can be visualized in relation to today's administrative framework. This method allows for a more coherent comparison of medical resource distribution over time, despite the extensive boundary changes.

First, we integrate historical data on physician density (outpatient physicians per 100,000 inhabitants) for four different time points and visualize them in one thematic map using the administrative boundaries valid at the respective times (see Fig. 4.15). These time points include 1884 during the German Empire, 1930 during the Weimar Republic, 1964 during the period of German division, and 2000 after reunification. Subsequently, the three older historical datasets will be transformed using the raster and zonal statistics method to align with the administrative boundaries of the year 2000. This will allow for a more coherent comparison through thematic maps, ensuring a clearer visualization of changes in physician density over time (see Fig. 4.16).

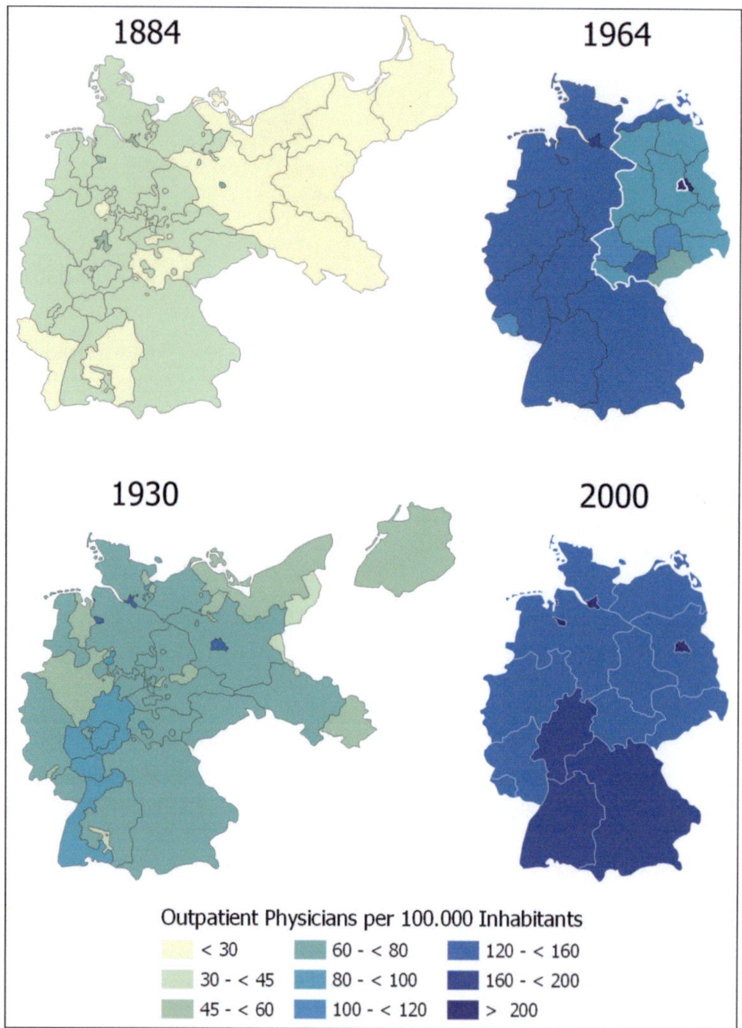

Fig. 4.15 Visualization of historical physician densities in Germany

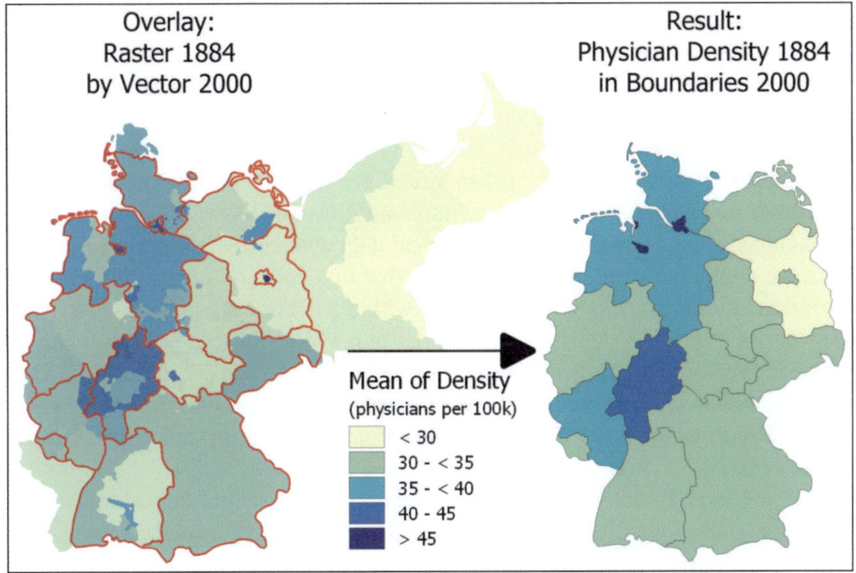

Fig. 4.16 Calculating the mean physician density of 1884 in the boundaries of 2000

The four-part thematic map in Fig. 4.15 illustrates the evolution of physician density across German regions for the years 1884, 1930, 1964, and 2000. A unified color gradient with nine classes, ranging from yellow to green to dark blue, was applied to ensure comparability across all four time points. This visualization effectively highlights the steady increase in physician density over the 116-year period.

As early as 1884, the city-states of Berlin, Hamburg, and Bremen already exhibited significantly higher physician densities compared to the larger territorial states. In contrast, the eastern provinces had the lowest densities, with only 17 physicians per 100,000 inhabitants in East Prussia and Posen. By 1930, this fundamental pattern remained, although physician density had risen considerably across all regions. The largest increases occurred in Hesse and Baden, while the city-states continued to stand out with the highest concentrations of physicians.

By 1964, the division of Germany into West Germany and the GDR had become evident not only politically but also in the distribution of healthcare resources. The western federal states displayed relatively homogeneous physician densities at a significantly higher level than those in the eastern districts. However, Hamburg and Berlin once again showed the highest physician densities, with East Berlin, despite being part of the GDR, maintaining a similar trend.

By 2000, physician densities had largely equalized across the reunified Germany. However, an important shift had taken place: the former West–East divide in physician density had transformed into a North–South divide. The introduction of physician needs-based planning during this period—intended to regulate new medical practices and contain rising healthcare costs—resulted in a slower increase in physician density overall. The most pronounced growth occurred in southern Germany, yet the city-states of Berlin, Hamburg, and Bremen continued to record the highest physician densities, much like at the beginning of the time series.

This analysis demonstrates that, while physician density has steadily increased across Germany over time, regional disparities have shifted rather than disappeared—evolving from an east–west imbalance into a north–south contrast.

To enhance the comparability of historical data with current administrative divisions, we apply a methodological approach that first converts vector polygons into raster data. In this process, each raster cell within a historical administrative unit is assigned the physician density value of that unit (Tool: Raster $\rightarrow$ Conversion $\rightarrow$ Rasterize (Vector to Raster)). The resulting 1884 physician density raster is shown in Fig. 4.16, with a continuous color gradient from yellow (lowest value $= 17$ physicians per 100,000 inhabitants) to dark blue (highest value $= 80$ physicians per 100,000 inhabitants).

Next, we overlay this raster with the vector data representing the administrative boundaries of the year 2000. Using the "Zonal Statistics" tool from the Raster Analysis Toolbox, we compute the arithmetic mean of all raster cells within each 2000 administrative unit. This results in an aggregated 1884 physician density (outpatient physicians per 100,000 inhabitants) estimate based on the 2000 boundaries, allowing for a direct comparison of historical and modern spatial distributions (see Fig. 4.16). Of course, it is important to acknowledge that this approach provides only a rough estimate. The physician density within a historical region was certainly not evenly distributed. However, the raster/vector method assumes uniformity, which can lead to distortions. For example, it is likely that, both historically and today, physician density in Brandenburg was higher in areas near the outskirts of Berlin than in the rest of the region. If this spatial variation was considered, Berlin would likely show an even higher value in the mean density map.

Using this method, we estimate physician densities for the three historical time points (1884, 1930, and 1964) within the administrative boundaries of the year 2000. Figure 4.17 shows the results, juxtaposing the historical data projected onto current boundaries for all four time points. To ensure comparability, a customized quantile classification method was applied to each map, ensuring that approximately the same number of federal states fall into each class. At first glance, it becomes evident that the West–East disparity in physician density has long-standing historical roots and remained largely consistent between 1884 and 2000—despite an overall increase in physician density across all regions. In 1884, the highest physician densities were found in the regions corresponding to today's northwestern federal states, significantly exceeding those in the south. However, this pattern reversed between 1884 and 1930. Since 1930, the highest physician densities have been consistently observed in the southern states of Bavaria, Baden-Württemberg, and Hesse.

The applied change detection approach demonstrates the potential of our historical administrative boundary data for analyzing long-term spatial developments. By transforming vector-based historical data into a raster format and aggregating values

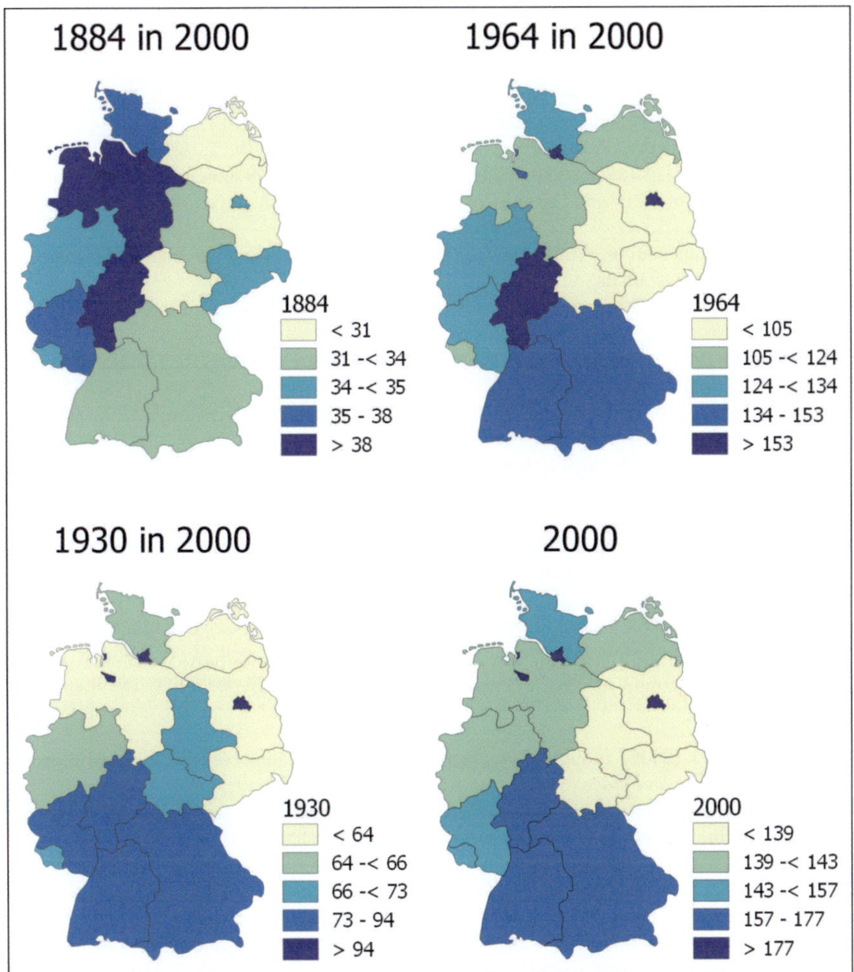

Fig. 4.17 Mean physician densities of historical censuses in the boundaries of 2000

to modern administrative units, we enable direct comparisons across different time periods despite shifting territorial boundaries.

This method not only reveals persistent regional disparities in physician density over more than a century but also highlights structural shifts, such as the emergence of higher densities in southern Germany. More broadly, it underscores the value of our dataset as a foundation for integrating historical statistics with contemporary spatial frameworks, facilitating diachronic analyses across diverse research fields.

References

Central Statistics Office (Ireland): Historical Reports. (2025). Online (05.05.2025). https://www.cso.ie/en/census/censusvolumes1926to1991/historicalreports/

Doetsch, J. N., Almendra, R., Severo, M., Leão, T., Pilot, E., Krafft, T., & Barros, H. (2023). 2008 economic crisis impact on perinatal and infant mortality in Southern European countries. *Journal of Epidemiology and Community Health, 77*(5), 305–314. https://doi.org/10.1136/jech-2022-219639

Esri ArcGIS Documentation. (2025). Online (05.05.2025). https://www.esri.com/en-us/arcgis/products/arcgis-pro/resources

Eurostat. (ed.). (2004). *Regionen: Systematik der Gebietseinheiten für die Statistik NUTS-2003.* Luxemburg

Eurostat. (ed.). (2025a). *Infant mortality by NUTS 2 region.* https://doi.org/10.2908/demo_r_minf. Last update: 03/03/2025.

Eurostat. (ed.). (2025b). *Infant mortality rates by NUTS 2 region.* https://doi.org/10.2908/demo_r_minfind. Last update: 14/03/2025.

Eurostat. (ed.). (2025c). *Statistics explained. Mortality and life expectancy statistics.* Online (05.05.2025). https://ec.europa.eu/eurostat/statistics-explained/index.php?title=Category:Mortality_and_life_expectancy

Gonzalez, R. M., & Gilleskie, D. (2017). Infant mortality rate as a measure of a country's health: A robust method to improve reliability and comparability. *Demography, 54*(2), 701–720. https://doi.org/10.1007/s13524-017-0553-7

GRASS GIS Documentation. (2025). Online (05.05.2025). https://grass.osgeo.org/documentation/

Harpur, A., Minton, J., Ramsay, J., McCartney, G., Fenton, L., Campbell, H., & Wood, R. (2021). Trends in infant mortality and stillbirth rates in Scotland by socio-economic position, 2000–2018: A longitudinal ecological study. *BMC Public Health, 21*(1), 995. https://doi.org/10.1186/s12889-021-10928-0

Marti-Henneberg, J. (2005). Empirical evidence of regional population concentration in Europe. In *Population, space and place* (Vol. 11, no. 4, pp. 269–281). Chichester.

Marti-Henneberg, J. (ed.). (2021). *European Regions, 1870—2020. A geographic and historical insight into the process of European integration.* Springer. https://doi.org/10.1007/978-3-030-61537-6

Onambele, L., San Martin-Rodríguez, L., Niu, H., Alvarez-Alvarez, I., Arnedo-Pena, A., Guillen-Grima, F., & Aguinaga-Ontoso, I. (2019). Mortalidad infantil en la Unión Europea: Análisis de tendencias en el período 1994–2015 [Infant mortality in the European Union: A time trend analysis of the 1994–2015 period]. *Anales De Pediatria, 91*(4), 219–227. https://doi.org/10.1016/j.anpedi.2018.10.022

QGIS Documentation. (2025). Online (05.05.2025). https://docs.qgis.org/

QGIS Download. (2025). Online (05.05.2025). https://qgis.org/download/

R Manuals. (2025). Online (05.05.2025). https://cran.r-project.org/manuals.html

Reidpath, D. D., & Allotey, P. (2003). Infant mortality rate as an indicator of population health. *Journal of Epidemiology and Community Health, 57*(5), 344–346. https://doi.org/10.1136/jech.57.5.344

Schön, L. (2007). *Sweden—Economic growth and structural change, 1800–2000.* EH.Net Encyclopedia. Online (05.05.2025). https://eh.net/encyclopedia/sweden-economic-growth-and-structural-change-1800-2000/

Simple Feature for R. https://r-spatial.github.io/sf/

Wolfe, B. L. (1986). Health status and medical expenditures: Is there a link? *Social Science & Medicine (1982), 22*(10), 993–999. https://doi.org/10.1016/0277-9536(86)90199-1

Zilidis, C., & Hadjichristodoulou, C. (2020). Economic crisis impact and social determinants of perinatal outcomes and infant mortality in Greece. *International Journal of Environmental Research and Public Health, 17*(18), 6606. https://doi.org/10.3390/ijerph17186606

Chapter 5
Conclusion

This work demonstrates the substantial potential of digitally modeling historical administrative boundaries for historical–geographical research. The development and application of the Regional European Geographical Information System (REGIS) and Regional European Thematic Mapping (REThM) databases provide a solid foundation for the systematic analysis of territorial changes in Europe between 1870 and 2020. By employing a precise, vector-based mapping of administrative units in a consistent and interoperable data format (GeoPackage), these resources enable both diachronic and comparative analyses across various spatial scales.

The integration of historical data allows for in-depth analysis of socioeconomic and demographic developments in their historical context. The standardized modeling of border geometries—supported by GIS methods for generalization, geometry unification, and attribute linking—creates an analytically robust basis for both quantitative and qualitative evaluations. In particular, the ability to compare historical and modern administrative units represents a significant methodological advancement.

Nonetheless, the methodological framework also reveals several challenges that must be addressed in future work. One major concern is the pragmatic decision to register boundary changes only when an area alteration exceeds a 10% threshold. Although this simplifies comparisons on a macro level, it carries the risk of overlooking smaller yet potentially significant administrative or politically relevant changes. Additionally, the quality of the underlying map sources is crucial, as a tension often exists between preserving historical authenticity and achieving technical generalization.

Furthermore, the manual post-processing of digitized boundary geometries, although precise, is labor-intensive and difficult to scale. For a comprehensive historical representation of European territories over a span of 150 years, automated methods for detecting and transforming boundary changes will be indispensable in the future. The application of machine learning algorithms and deep learning approaches for boundary recognition in scanned maps, combined with natural

© The Author(s) 2026

81

J. Schweikart et al., *Regional European Geographic Information System (REGIS),
1870–2020*, SpringerBriefs in Geography, https://doi.org/10.1007/978-3-032-14932-9_5

language processing for extracting historical texts and census data, offers promising potential for significant progress in this area.

Future developments should focus on further advancing the REGIS system as an open, collaborative research tool. Key tasks include the systematic integration of additional geometry and attribute data, the inclusion of comprehensive metadata regarding the source, accuracy, and legal validity of individual boundary lines, and the development of user-friendly interfaces that broaden accessibility beyond the GIS community. Moreover, increasing the interoperability with external databases—especially in demography, economic history, and environmental research—opens new opportunities for interdisciplinary research endeavors.

The application of the REGIS and REThM databases within the QGIS environment illustrates how digital tools can contribute to the analysis of historical administrative structures. The possibility of combining external datasets, reclassifying information via quantile-based methods, and comparing historical statistics against standardized modern administrative units marks a significant breakthrough in historical GIS. This standardization is key to overcoming the long-standing problem of fragmented data within European studies, enabling coherent comparisons over time and space.

At the same time, the results underscore the need for continuous review and refinement of the methodological framework. In particular, for small-scale analyses, generalization processes remain vulnerable to losses in precision. Achieving a balance between visual consistency and content accuracy requires an ongoing interplay between automated and manual procedures—a challenge that is especially pronounced in regions with incompletely documented administrative developments or heterogeneous source material. Here, future algorithmic improvements, especially in harmonizing differently processed geodata, hold considerable promise.

In summary, REGIS makes a substantial contribution to the spatio-temporal analysis of Europe's territorial evolution. The provision of scalable, standardized data formats establishes a robust foundation for historical research as well as for political and social analyses. The databases not only offer new perspectives for visualizing complex historical processes but also support the empirical testing of theoretical assumptions regarding state transformation, regionalization, and integration. Their utility extends beyond academic research to applications in education, public knowledge dissemination, and policy planning with historical context.

The potential of this digital infrastructure remains far from exhausted. Future efforts should increasingly focus on automating currently manual processes, enhancing interoperability with thematically related data sources, and integrating uncertainties and ambiguities into historical map products. In this way, REGIS and REThM function not only as retrospective documentation tools but also as dynamic platforms for the further development of historical spatial analysis.

Appendix

Disclaimer and Legal Notice Regarding the Use of REGIS/ REThM Data

The spatial data presented in the REGIS database are the result of a comprehensive and carefully executed digitization process of historical administrative boundaries in Europe between 1870 and 2020. These data were derived from a wide range of historical cartographic sources, including maps and atlases. Most of them are in the public domain due to their age or have been made freely available through academic and archival institutions. In addition, we have made use of freely accessible and appropriately licensed online sources, such as CAI Maps, the University of Texas Map Collection, Wikimedia Maps, Wikipedia Maps, and the David Rumsey Map Collection. Use of the Eurostat reference dataset from 1990 was authorized for non-commercial purposes.

It is important to note that no original map lines were directly reproduced in REGIS. Every boundary line was independently reconstructed and generalized by manually placing control points in a Geographic Information System (GIS). The REGIS dataset therefore represents a new and original work, derived through digitization and reinterpretation of historical maps—many of which are over a century old or in the public domain. All adopted boundaries were manually generalized and geometrically adjusted using GIS software, often with only small sections of boundaries extracted from larger multi-thematic maps. This process resulted in the creation of a new and original dataset which, while based on historical sources, constitutes a distinct intellectual and cartographic work.

The authors and publishers of this dataset assume no liability for any direct or indirect damage, losses, or claims arising from the use, reuse, or interpretation of the data. Users are solely responsible for ensuring that their use of the data complies with relevant legal, ethical, and academic standards, especially in cases of redistribution, publication, or derivative work. We recommend that users independently verify the legal status of individual source materials when undertaking further analysis or publication based on the REGIS data.

© The Editor(s) (if applicable) and The Author(s) 2026 85
J. Schweikart et al., *Regional European Geographic Information System (REGIS),
1870–2020*, SpringerBriefs in Geography, https://doi.org/10.1007/978-3-032-14932-9

The historical maps used in this project, including those without explicitly stated authorship or publication information, were sourced from secondary compilations of administrative-territorial developments (e.g., atlases or scanned documents). Despite careful efforts, in some cases it was not possible to identify the original creator, publisher, or copyright holder. The use of these materials is strictly limited to academic, non-commercial purposes, and all data derived from such maps has been independently digitized and generalized using GIS techniques.

Funded by the European Union. Views and opinions expressed are however those of the author(s) only and do not necessarily reflect those of the European Union or the European Education and Culture Executive Agency (EACEA). Neither the European Union nor EACEA can be held responsible for them.

License and Terms of Use

The REGIS/REThM dataset is released under the terms of the Creative Commons Attribution-NonCommercial 4.0 International License (CC BY-NC 4.0).

This means that the data may be freely used, shared, and adapted for non-commercial, academic, or educational purposes, provided that:

- Proper attribution is given to the creators, citing this publication as the original source.
- Any modifications, derivatives, or secondary outputs clearly indicate how the original dataset has been altered.

Citation Recommendation

Schweikart, J., Martí-Henneberg, J., Pieper, J., Morillas-Torné, M. & Schulte, B. (202x): Regional European Geographic Information System (REGIS), 1870 – 2020. A GIS database for thematic mapping and analytical purpose. Springer Nature.

Non-commercial Use Clause

This dataset is for scientific, research, and educational purposes. Any use of the dataset for commercial purposes (e.g., in for-profit products, consultancy services, or monetized platforms) is explicitly prohibited without prior written consent from the authors or the hosting institution.

Disclaimer Regarding Source Materials

The authors have taken every possible measure to ensure that the database was developed in strict accordance with scientific standards and with full transparency.

Several historical maps used for reconstructing administrative boundaries were sourced from the **David Rumsey Map Collection**, hosted by the David Rumsey Map Center at Stanford Libraries. These maps were used strictly as visual references for boundary digitization in GIS and were not reproduced in full or used as stand-alone publications.

According to David Rumsey Collection's usage guidelines, images may be used at no charge for personal use or inclusion in print and digital publications. Since the maps were employed to produce highly derivative boundary data and not reproduced directly, their use falls within the permitted scope. For maps published after 1929, we made good-faith efforts to assess copyright status and believe our academic, non-commercial, and derivative use to fall under fair use principles.

All cited maps from the Rumsey Collection include full bibliographic references, URLs, and credit as requested:

"David Rumsey Map Collection, David Rumsey Map Center, Stanford Libraries" (see https://www.davidrumsey.com/about/copyright-and-permissions).

The maps created by the **U.S. Central Intelligence Agency (CIA)** used in this project are in the public domain, as they are works of the US Federal Government. Access to these maps was facilitated through the Perry-Castañeda Library Map Collection (PCLMC) hosted by the University of Texas at Austin. While the PCLMC provides open access to these materials, it does not hold the copyright nor act as the original publisher. Source: **Perry-Castañeda Library Map Collection, University of Texas at Austin** (https://legacy.lib.utexas.edu/maps/).

The maps retrieved from **Wikimedia Commons** and **Wikipedia** were used based on their availability under open licenses (typically Creative Commons or public domain). The authors and uploaders of these maps are credited on the respective Wikimedia pages. License terms may vary and are subject to change; therefore, users are encouraged to consult the specific file pages for up-to-date copyright and reuse information. Inclusion of these maps in this work does not imply endorsement by Wikimedia Foundation or the original contributors. Source: Wikimedia Commons and Wikipedia (https://commons.wikimedia.org, https://www.wikipedia.org).

Map Sources

Andree, R.; Scobel, A. (1906): Andrees Allgemeiner Handatlas. 5th Edition. Velhagen & Klasing, Bielefeld. [used for: Czechoslovakia, Denmark, Greece, Italy, Memel Territory, Montenegro, Ottoman Empire, Portugal, Romania, Serbia, England, Ireland, Northern Ireland, Scotland, Wales]

Appleton, D. & Co. (1892): The Library Atlas of Modern Geography. Switzerland 19. D. Appleton & Co., New York. Online (05.05.2025): https://www.davidrumsey.com/luna/servlet/detail/RUMSEY~8~1~260851~5523047 David Rumsey Map Collection, David Rumsey Map Center, Stanford Libraries. [used for: Switzerland]

Bartholomew, J. (1955): The Times Atlas of the World. Mid-Century Edition, Plate 53: Denmark – Faeroes – Norway West. Vol. III. Houghton Mifflin Co., John Bartholomew & Son Ltd., Boston/London. Online (05.05.2025): https://www.davidrumsey.com/luna/servlet/detail/RUMSEY~8~1~225639~5506111 David Rumsey Map Collection, David Rumsey Map Center, Stanford Libraries. [used for: Denmark]

Bartholomew, J.G. (1922): Times Survey Atlas of the World – Germany Southern Section. The Times, London. Online (05.05.2025): https://www.davidrumsey.com/luna/servlet/detail/RUMSEY~8~1~3128~440030 David Rumsey Map Collection, David Rumsey Map Center, Stanford Libraries. [used for: Saarland]

Bartholomew, J. G. (1904): Handy Reference Atlas of the World. Maps: "Ireland Northern Section" and "Ireland Southern Section". [used for: Ireland]

Bayer, H. (1953): World Geographic Atlas: A Composite of Man's Environment – Central Europe. Container Corporation of America, Chicago. Online (05.05.2025): https://www.davidrumsey.com/luna/servlet/detail/RUMSEY~8~1~217962~5503996 David Rumsey Map Collection, David Rumsey Map Center, Stanford Libraries. [used for: Poland]

Bohinec, V.; Flanina, F. (1958): FLR Jugoslavija. Avtokarta. Avto-Moto Zveza Slovenije, Ljubljana. [used for: Yugoslavia]

Cartografisch Instituut P. Mantnieks (197-?): Latvija. [used for: Latvia]

Directia Topografică Militară (1960): Republica Populară Romînă – Harta Politico-Administrativă. Direcția Topografică Militară, Bucharest. [used for: Romania]

Generalstab des Heeres, Abt. f. Kriegskarten u. Vermessungswesen (1940): Rumänien / Verwaltungsübersicht. Generalstab des Heeres, Berlin. [used for: Romania]

Götz, A. (1966): Atlas Československé Socialistické Republiky – Development of Territorial Organisation. Československá akademie věd, Ústřední správa geodézie a kartografie, Prague. Online (05.05.2025): https://www.davidrumsey.com/luna/servlet/detail/RUMSEY~8~1~235033~5510246 David Rumsey Map Collection, David Rumsey Map Center, Stanford Libraries. [used for: Czechoslovakia]

Gray, O.W. (1873): Gray's Atlas Of The United States, With General Maps Of The World. Germany No. 1. Prussia and Saxony. Stedman, Brown & Lyon, Philadelphia. Online (05.05.2025): https://www.davidrumsey.com/luna/servlet/detail/RUMSEY~8~1~264413~5524825 David Rumsey Map Collection, David Rumsey Map Center, Stanford Libraries. [used for: Germany]

Haack, H. (1965): Deutsche Demokratische Republik – Bergbau und Industrie. Online (05.05.2025): https://www.davidrumsey.com/luna/servlet/detail/RUMSEY~8~1~289812~90061366 David Rumsey Map Collection, David Rumsey Map Center, Stanford Libraries. [used for: GDR]

Haack, H.; Stieler, A. (1925): Stieler's Atlas of Modern Geography – Composite: Italy. Justus Perthes, Gotha. Online (05.05.2025): https://www.davidrumsey.com/luna/servlet/detail/RUMSEY~8~1~281354~90054133 David Rumsey Map Collection, David Rumsey Map Center, Stanford Libraries. [used for: Italy]

Hammond, C.S. (1948): Hammond's New World Atlas – Finland, Poland, Austria, Czechoslovakia and Hungary, Yugoslavia. Garden City Publishing Company, Inc., Garden City. Online (05.05.2025): https://www.davidrumsey.com/luna/servlet/detail/RUMSEY~8~1~206037~3002874 David Rumsey Map Collection, David Rumsey Map Center, Stanford Libraries. [used for: Poland]

Holm Sundhausen (1989): Historische Statistik Serbiens 1834–1914. Mit europäischen Vergleichsdaten. Südosteuropäische Arbeiten, Band 87. Hrsg. von Mathias Bernath und Karl Nehring für das Südost-Institut München. R. Oldenbourg Verlag, München. [used for: Serbia]

Jāṇa Sēta (1996): Eesti Halduskaart. Jāṇa Sēta, Riga. [used for: Estonia]

Књижарница (1930): Карта Краљевине Југославије. Издавачка Књижарница Геце Кона Београд, Belgrade. [Used for: Yugoslavia]

Kowalski, Tadeusz; Ungar, T. (1939): This is Poland. Edward Stanford Ltd., London. [used for: Poland]

Маринковић, Влад. (1922): Карта Нове Админстративне Поделе Краљевине Срба, Хрвата И Словенаца. Свесловенске Књижаре, Belgrade. [used for: Yugoslavia]

Reichsamt für Landesaufnahme Berlin (1930): Amtliche Karte der Freien Reichsstadt (1919–1939). Online (05.05.2025): https://de.wikipedia.org/wiki/Datei:Karte_der_Freien_Stadt_Danzig_100k_1930.jpg [used for: Free City of Danzig]

Scobel, A (1906): Andree's General Hand Atlas. 5th edition. Velhagen & Klasing, Bielefeld.

Sohr, Karl; Berghaus, Heinrich; Handtke, Friedrich; Klein, H. J. (1888): Sohr-Berghaus Hand-Atlas über alle Theile der Erde, No. 69: Die Balkan-Halbinsel, Nordwestlicher Theil. C. Flemming, Glogau. Online (05.05.2025): https://www.davidrumsey.com/luna/servlet/detail/RUMSEY~8~1~302174~90073169:No69--Die-Balkan-Halbinsel-Nord-wes David Rumsey Map Collection, David Rumsey Map Center, Stanford Libraries. [used for: Serbia]

Statistical Office of Estonia (Statistikaamet) (1922). Proportion of Urban Population, 1922 (Linnarahvastiku osatähtsus, 1922). [used for: Estonia]

Stieler, Adolf (1882). Österreich-Ungarn (Composite map: Südwestlicher Blatt & Nordwestlicher Theil). Justus Perthes Verlag, Gotha. Online (05.05.2025): https://www.davidrumsey.com/luna/servlet/detail/RUMSEY~8~1~243050~5513273:Composite--Oster-reich_Ungarn,?sort=pub_list_no_initialsort%2Cpub_date%2Cpub_list_no%2Cseries_no David Rumsey Map Collection, David Rumsey Map Center, Stanford Libraries. [used for: Austria-Hungary]

Stieler, Adolf (1871). Länder der Ungarischen Krone. Justus Perthes Verlag, Gotha. Online (05.05.2025): https://www.davidrumsey.com/luna/servlet/det ail/RUMSEY~8~1~242135~5512888:Laender-der-Ungarischen-Krone--Un-ga David Rumsey Map Collection, David Rumsey Map Center, Stanford Libraries. [used for: Austria-Hungary and Hungary]

Stieler, Adolf; Domann, B. (1911). Stieler's Hand-Atlas, No. 51: Balkan-Halbinsel. Justus Perthes, Gotha. Online (05.05.2025): https://www.davidr umsey.com/luna/servlet/detail/RUMSEY~8~1~297337~90068929:Nr--51--Bal kan-Halbinsel,-Bl--1---- David Rumsey Map Collection, David Rumsey Map Center, Stanford Libraries. [used for: Serbia]

Stieler, Adolf; Vogel, C. (1911). Österreich-Ungarn in vier Blättern (Composite map: Nos. 17–20). Justus Perthes Verlag, Gotha. Online (05.05.2025): https:// www.davidrumsey.com/luna/servlet/detail/RUMSEY~8~1~297270~90068862: Composite-Map--Nr--17-20--Oester-rei David Rumsey Map Collection, David Rumsey Map Center, Stanford Libraries. [used for: Austria-Hungary]

Stieler, Adolf; Vogel, C. (1873). Stieler's Handatlas über alle Theile der Erde und über das Weltengebäude. Composite: Deutschland. Justus Perthes, Gotha. Online (05.05.2025): https://www.davidrumsey.com/luna/servlet/detail/RUM SEY~8~1~297270~90068862:Composite-Map--Nr--17-20--Oester-rei David Rumsey Map Collection, David Rumsey Map Center, Stanford Libraries. [used for: Germany]

Touring Club Italiano (1929). Atlante internazionale del Touring Club Italiano. Milano: Touring Club Italiano. [used for: Germany, Italy, Norway, Poland]

Unknown author (n.d.): Административно-териториално устройство на България (1880–1934). Composite historical maps of administrative-territorial division. Language: Bulgarian. Source: scanned image from unidentified secondary publication. [used for: Bulgaria]

U.S. Central Intelligence Agency (2002). Lithuania (Political). Online (05.05.2025): https://maps.lib.utexas.edu/maps/europe/lithuania_pol_2002. jpg University of Texas Libraries [used for: Lithuania]

U.S. Central Intelligence Agency (2000). Albania (Political). Online (05.05.2025): https://maps.lib.utexas.edu/maps/europe/albania_pol00.jpg University of Texas Libraries [used for: Albania]

U.S. Central Intelligence Agency (2000). Croatia (Political). Online (05.05.2025): https://maps.lib.utexas.edu/maps/europe/croatia_rel_00.jpg University of Texas Libraries [used for: Croatia]

U.S. Central Intelligence Agency (2000). Poland (Political). Online (05.05.2025): https://maps.lib.utexas.edu/maps/europe/poland_rel00.jpg University of Texas Libraries [used for: Poland]

U.S. Central Intelligence Agency (1998). Latvia (Political). Online (05.05.2025): https://maps.lib.utexas.edu/maps/europe/latvia_pol98.jpg University of Texas Libraries [used for: Latvia]

U.S. Central Intelligence Agency (1996). Romania (Political). Online (05.05.2025): https://maps.lib.utexas.edu/maps/europe/romania_pol96.jpg University of Texas Libraries [used for: Romania]

U.S. Central Intelligence Agency (1994). Czech Republic (Political). Online (05.05.2025): https://maps.lib.utexas.edu/maps/europe/czechrepublic.jpg University of Texas Libraries [used for: Czechia, Czechoslovakia]

U.S. Central Intelligence Agency (1994). Slovakia (Political). Online (05.05.2025): https://maps.lib.utexas.edu/maps/europe/slovakia_map.jpg University of Texas Libraries [used for: Slovakia, Czechoslovakia]

U.S. Central Intelligence Agency (1994). Hungary (Political). Online (05.05.2025): https://maps.lib.utexas.edu/maps/europe/hungary.jpg University of Texas Libraries [used for: Hungary]

U.S. Central Intelligence Agency (1987). Northern Ireland (UK) (Political). Online (05.05.2025): https://maps.lib.utexas.edu/maps/europe/northern_ireland_pol87.jpg University of Texas Libraries [used for: Northern Ireland]

U.S. Central Intelligence Agency (1982). Poland (Political). Online (05.05.2025): https://maps.lib.utexas.edu/maps/europe/poland.jpg University of Texas Libraries [used for: Poland]

Vivien de Saint-Martin & Schrader (1937): Atlas Universel de Géographie. Europe Orientale, Feuille Nord. Librairie Hachette, Paris. Online (05.05.2025): https://www.davidrumsey.com/luna/servlet/detail/RUMSEY~8~1~36531~120 1086:Europe-Orientale,-Feuille-Nord-?sort=pub_list_no_initialsort%2Cpub_date%2Cpub_list_no%2Cseries_no&qvq=w4s:/when%2F1937;q:Finland;sort: pub_list_no_initialsort%2Cpub_date%2Cpub_list_no%2Cseries_no;lc:RUM SEY~8~1&mi=0&trs=1# David Rumsey Map Collection, David Rumsey Map Center, Stanford Libraries. [used for: Finland]

Wikimedia Commons (2000): Administrative Division Bulgaria 1951–1959. Online (05.05.2025): https://commons.wikimedia.org/wiki/File:Administrative_Division_Bulgaria_1951-1959.jpg [used for: Bulgaria]

Wikimedia Commons (2014): LithuaniaCounties1918-1940.png. Online (05.05.2025): https://commons.wikimedia.org/wiki/File:LithuaniaCounties1 918-1940.png [used for: Lithuania]

Wikimedia Commons (2013): Statistical regions of Serbia.svg. Online (05.05.2025): https://commons.wikimedia.org/wiki/File:Statistical_regions_of_Serbia.svg [used for: Yugoslavia]

Wikimedia Commons (2012): Slovenia, administrative divisions – Nmbrs (statistical regions) – colored.svg. Online (05.05.2025): https://commons.wikimedia.org/wiki/File:Slovenia,_administrative_divisions_-_Nmbrs_(statistical_region s)_-_colored.svg [used for: Slovenia]

Wikimedia Commons (2011): Regions of Bulgaria Map.png. Online (05.05.2025): https://commons.wikimedia.org/wiki/File:Regions_of_Bulgaria_Map.png [used for: Bulgaria]

Wikimedia Commons (2008): ScotlandLabelled.png. Online (05.05.2025): https://commons.wikimedia.org/wiki/File:ScotlandLabelled.png [used for: Scotland]

Wikimedia Commons (2007): Subcarpathia Carpatho-Ukraine german.svg. Online (05.05.2025): https://commons.wikimedia.org/wiki/File:Subcarpathia_Carpatho-Ukraine_german.svg [used for: Czechoslovakia]

Wikimedia Commons (2006): Freistaat Fiume 1920–1924.png. Online (05.05.2025): https://commons.wikimedia.org/wiki/File:Freistaat_Fiume_1920-1924.png [used for: Free State of Fiume]

Wikimedia Commons (2005): Territorio libero di Trieste carta.png. Online (05.05.2025): https://commons.wikimedia.org/wiki/File:Territorio_libero_di_Trieste_carta.png [used for: Free Territory of Trieste]

Wikipedia (2013): Czech Republic, administrative divisions – de++ – colored.svg. Online (05.05.2025): https://de.wikipedia.org/wiki/Datei:Czech_Republic,_administrative_divisions_-_de%2B%2B_-_colored.svg [used for: Czechia]

Wikipedia (2010): Iceland regions.svg. Online (05.05.2025): https://de.wikipedia.org/wiki/Datei:Iceland_regions.svg [used for: Iceland]

Wikipedia (2009): Bosnia and Herzegovina Political.png. Online (05.05.2025): https://de.wikipedia.org/wiki/Datei:Bosnia_and_Herzegovina_Political.png [used for: Bosnia-Herzegovina]

Wikipedia (2008): Map Cymru 1996 gyda rhifau.svg. Online (05.05.2025): https://en.wikipedia.org/wiki/File:Map_Cymru_1996_gyda_rhifau.svg [used for: Wales]

Wikipedia (2007): Kraj slovakia german.svg. Online (05.05.2025): https://de.wikipedia.org/wiki/Datei:Kraj_slovakia_german.svg [used for: Slovakia]